Adventures in Mathematics

Adventures in Mathematics

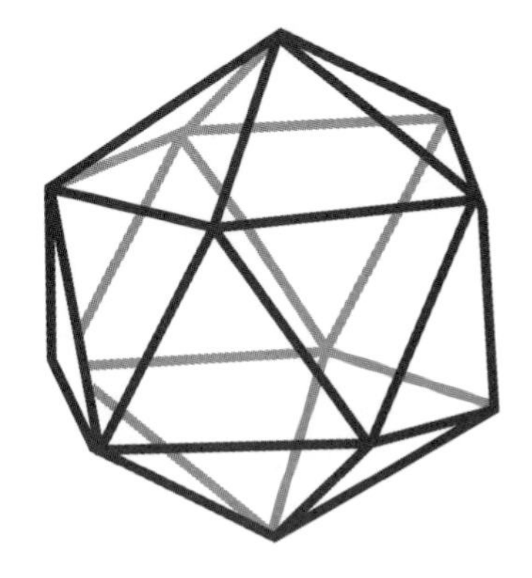

Martin A. Moskowitz
with the assistance of Hossein Abbaspour
City University of New York

NEW JERSEY • LONDON • SINGAPORE • SHANGHAI • HONG KONG • TAIPEI • BANGALORE

Published by

World Scientific Publishing Co. Pte. Ltd.
5 Toh Tuck Link, Singapore 596224
USA office: Suite 202, 1060 Main Street, River Edge, NJ 07661
UK office: 57 Shelton Street, Covent Garden, London WC2H 9HE

British Library Cataloguing-in-Publication Data
A catalogue record for this book is available from the British Library.

ADVENTURES IN MATHEMATICS

ISBN 981-238-683-1 (pbk)

Printed in Singapore by World Scientific Printers (S) Pte Ltd

Preface

Our purpose here is to present mathematics as it really is: a vast panorama of ideas, whose history reflects some of the most noble thoughts of mankind. This book represents an attempt, although on a limited scale, to convey some of this to the reader who perhaps has little technical or mathematical background, but who is truly interested in ideas and understanding certain aspects of the world around us. The material would also be appropriate to high school or community college teachers who seek an overview of mathematics so they can see through the formalities of the standard curriculum and, in turn, convey to their students some of the real scope and power to be found in mathematical ideas. Selections from the subject matter presented here would also be very suitable for a college pre-calculus course. Indeed, one of our objectives is to try to find and to understand some real mathematics without the reader needing to have a knowledge of calculus.

It is sometimes harmful to a beginner to be overly burdened with the usual standards of rigorous proof employed by the professional mathematician. In fact, it is often helpful to let geometric intuition be a guide to reasoning about certain of these matters. This will be the attitude adopted in this book. When it is helpful to give a rigorous proof we will; when it is more useful to do otherwise we will do that, it being understood that outlines of proofs can be filled in and made completely rigorous. From time to time, to give an idea of the horizons, we will state a result without proof.

As can be seen from the table of contents this book deals with numbers (and in reality some number theory), groups (and their actions as transformation groups) as well as linear algebra, geometry and topology in low dimensions. It is the author's hope that the material contained between these covers represents much that is central to mathematics and/or offers useful and important models for thinking about our subject. But it also must be admitted that, to some degree, the subject matter reflects some of the author's interests and perhaps even prejudices. In any case, it is hoped that an instructor using this book will find ample material from which to choose his or her own preferred topics, perhaps occasionally supplementing the material with other sources, some of which are listed below.

Martin Moskowitz,
Spring 2003

For collateral or further reading the following books are recommended:

1. *What is Mathematics?*, Courant and Robbins, Oxford University Press, 1978

2. *Invitation to Mathematics*, Konrad Jacobs, Princeton University Press, 1992

3. *Fundamentals of Mathematics*, Moses Richardson, Macmillan and Co., 1966

4. *A Concise History of Mathematics*, Dirk J. Struik, Dover Books, 1966

5. *Symmetry*, Hermann Weyl, Princeton University Press, 1952

Notation

$\mathbb{Z}$ stands for the set of integers,

$\mathbb{Z}^+$ stands for the set of positive integers,

$\mathbb{Q}$ stands for the set of rational numbers,

$\mathbb{R}$ =the set of real numbers,

$\mathbb{C}$ =the set of complex numbers,

$[a, b] = \{x \in \mathbb{R} : \quad a \leq x \leq b\}$,

$[a, b) = \{x \in \mathbb{R} : \quad a \leq x < b\}$,

$(a, b] = \{x \in \mathbb{R} : \quad a < x \leq b\}$,

$(a, b) = \{x \in \mathbb{R} : \quad a < x < b\}$.

Σ will stand for the sum of a finite number of items and Π will stand for their product.

If A is a set and a is an element of it, we shall write $a \in A$.

If A and B are sets and every element of A is an element of B, we shall write $A \subseteq B$, to be read A is a *subset* of B.

$A = B$ means $A \subseteq B$ and $B \subseteq A$. That is A and B consist of exactly the same elements.

If A and B are sets then $A \cup B$, the *union* of A and B, is the set of those elements in either A or B, while $A \cap B$, the *intersection* of A and B, is the set of elements in both A and B.

If $B \subseteq A$ then we shall write $A \setminus B$ to indicate the things in A, but not in B

As is customary, the end of a proof will be indicated a small square box $\square$.

Contents

Chapter 1

What is a Number?

We begin with the basic properties of the number system and some generalizations. The concept of number is fundamental to mathematics and all that will follow.

Can you imagine a world without numbers? We all take them for granted, like air, water and blue skies. Yet they did not always exist; they were invented by human beings, although some mathematicians may feel that they were not invented, but rather discovered–that indeed they do exist "out there" like air, water and blue skies! The fact remains that numbers became a part of our lives around 5000 years ago, because they satisfied certain requirements. The most obvious, of course, was the need to measure quantities, such as time (in the devising of calendars), land for irrigation or after the annual flooding of a river, the amount of seed needed to cultivate a given field, or the quantity and size of stone required to build a pyramid. Such elementary and practical functions soon were found to possess an appeal that had not only to do with practical needs, i.e., the laws that seemed to govern numbers could be understood as a system that was not only rational, but had great aesthetic appeal. (It might be mentioned that both music and painting originated in ritual intended to produce a desired effect, and in this sense these human activities were also invented to serve practical needs such as propitiating the gods, thus ensuring adequate rainfall or a successful hunt; only relatively recently have art and music been asked to serve essentially and often exclusively aesthetic needs.) Numbers, then, satisfy at least three very essential human requirements: they serve the practical function of enabling us to measure or compute quantities (which may not be so easy to measure); they exert a strong appeal to the human faculty of rationality; and they form a system that is aesthetically appealing, thus satisfying our need for beauty.

But *what exactly is a number?* While we may all use numbers and have an intuitive sense of their definition, our purpose here will be to arrive at a more profound understanding of the basic properties of the number system. The number system is used not only throughout mathematics, and serves as its foundation, but also underlies, although sometimes only implicitly, virtually every area of human activity, including such fields as diverse as physics, chemistry, engineering, biology, the medical sciences, and music. All these modern disciplines, not to speak of medieval architecture and Renaissance painting, to name but a few, could not be done without numbers entering in significant ways.

We shall start with the first numbers to be invented, the ones that need nothing more than fingers and the like, to keep track of them. (For example, for Inuit peoples the largest number was 20, which was called "a man counted to the end"). These are the positive integers. After studying their most important property we shall see that there are some requirements which they do not fulfill. To put it another way, there are questions which can be posed in terms of positive integers alone, but which cannot be answered within that context. For this reason it will be necessary to turn to the *set* of all the integers.[1] At first glance these will seem to be sufficient, but we will discover that in fact the same situation obtains here, as indeed it will at every stage save the last. For this reason we will pass successively to the rational numbers, the real numbers and, finally, the complex numbers. Along the way it will be efficient to also consider polynomials and similar "number" systems. Later we will see how all this is connected with geometry, calculus and perhaps some other subjects as well.

[1]We take the notion of set, in an intuitive way, to mean any well defined collection of objects.

1.1 The Positive Integers

The *positive integers*, $\mathbb{Z}^+$, consist of the whole numbers beginning with 1. Thus

$$\mathbb{Z}^+ = \{1, 2, 3, \ldots\}$$

This set moves off to the right, or in the *positive* direction. When an integer m is to the right of another integer n, we shall say that n is less than m; in symbols, $n < m$. If we want to include the possibility that $n = m$ then we write $n \leq m$. Clearly, this set has the property that 1 is in it (written $1 \in \mathbb{Z}^+$) and if anything is in it, then so is the next thing (if $n \in \mathbb{Z}^+$, then also $n+1 \in \mathbb{Z}^+$). From this it follows that $\mathbb{Z}^+$ is an unbounded set, i.e., moves off infinitely far to the right. Since we are looking at infinity here, we are already at a higher level of abstraction than our early predecessors. It is also intuitively obvious from this last fact that the sum of any two numbers in $\mathbb{Z}^+$ is again in $\mathbb{Z}^+$. When a set such as $\mathbb{Z}^+$ has this property we shall say that it is *closed under addition*. Now, how can we convince ourselves that this last statement is true? Let m and n be the positive integers. Since $m \in \mathbb{Z}^+$, then so is m+1. But by the same reasoning so is $m + 2$. Continuing in this way (here one usually writes $\ldots$) we eventually come to $n + m$. Therefore $m + n \in \mathbb{Z}^+$.

There is a principle behind this trivial argument or *proof* which is worth isolating. Let S be any set of positive integers (written $S \subseteq \mathbb{Z}^+$). What do we need to know to be sure that S is all of $\mathbb{Z}^+$ (written $S = \mathbb{Z}^+$)? Clearly, we would have to know that $1 \in S$ and also that if anything, say $n \in S$, then so is $n+1 \in S$. This is because $\mathbb{Z}^+$ has these very properties! The question is *would this be enough*, or in technical language would it be *sufficient*? The answer to this question is yes and the reason is this. Suppose that S were not all of $\mathbb{Z}^+$. Then there would be something in $\mathbb{Z}^+$, not in S. But then, there would have to be a least such thing (furthest to the left), say n_0. Now $n_0 \neq 1$, since $1 \in S$. Look at the number immediately to the left of n_0. This number, say n, must be in S. But then $n + 1 = n_0$ must also be $\in S$. This gives a contradiction and so our original assumption that S is not all of $\mathbb{Z}^+$ is false. This is an example of what is called an *indirect proof* or *proof by contradiction*. We have just proven the *Principle of Induction* .

Theorem 1.1.1. *Suppose $S \subseteq \mathbb{Z}^+$ and has the properties that*

1. *$1 \in S$ and*

2. *For any $n \in \mathbb{Z}^+$ if $n \in S$, then $n + 1 \in S$.*

Then $S = \mathbb{Z}^+$.

Sometimes the principle that lies behind the Principle of Induction, namely that *any non empty set of positive integers has a least element*, is also called the Principle of Induction. The usual symbol for the *empty set* is the Greek letter ϕ. It is a technical convenience and means the set without any elements at all.

We can use either of these principles of induction to give a proof of the fact that $\mathbb{Z}^+$ is closed under addition. In this case it is more convenient to use the principle that each non empty set of positive integers has a least element. Let m and n be the positive integers. If $m+n$ is not in $\mathbb{Z}^+$, let k be the smallest positive integer such that $m+k$ is not in $\mathbb{Z}^+$. As we have seen, $m + 1 \in \mathbb{Z}^+$. Therefore $1 < k$. Since $k-1 < k$, $m+(k-1) \in \mathbb{Z}^+$. But adding 1 to a positive integer gives us another one (since this is how $\mathbb{Z}^+$ was constructed in the first place!). Thus, $m + (k-1) + 1 = m + k \in \mathbb{Z}^+$, a contradiction.

Another, perhaps more interesting, application of the principle of induction using a *direct argument* is in summing an *arithmetic progression*. Let n_0 and $d \in \mathbb{Z}^+$ and form the following integers:

$$n_0, n_0 + d, n_0 + 2d, \ldots \tag{1.1}$$

where kd means $d + d + \ldots$ (adding d to itself k times).

Notice that since $(k + l)d$ means adding d to itself $k + l$ times and this is clearly the same as adding d to itself k times, and then adding d to itself l times, we get for any three positive integers k, l and d that $(k+l)d = k \cdot d + l \cdot d$. This is called the *distributive law* and will be used in the future whenever needed.

A sequence given by (1.1) is called an *arithmetic progression* beginning with n_0 and with difference d. For example $\mathbb{Z}^+$ itself is the arithmetic progression beginning with 1 and with difference 1. Now, is there some convenient formula for the sum of the first $k+1$ terms of an arithmetic progression?

$$n_0 + n_0 + d + n_0 + 2d + \ldots + n_0 + kd$$

Clearly, this sum equals $(k+1)n_0 + d(1+2+\ldots k)$. Thus, answering this question comes down to calculating the sum (usually written Σ), $\Sigma(k) = 1+2+\ldots k$, of the first k positive integers. We will show that $\Sigma(k) = \frac{k(k+1)}{2}$.

Something about this formula requires an explanation. Namely, the right hand side looks as though it might not be an integer. But since, as we saw, the sum of any two integers is again an integer, the same is true for the sum of any finite number of integers.The reason for this is that even though we are dividing by 2 we still get an integer on the right side because either k or $k+1$ is even (divisible by 2).

To prove that this formula is correct we need only show that it is true when $k = 1$, i.e., that $1 = 1 \cdot 2/2$, an evident truth and that if $\Sigma(k) = \frac{k(k+1)}{2}$, then $\Sigma(k+1) = \frac{(k+1)(k+2)}{2}$. Now clearly, $\Sigma(k+1) = \Sigma(k) + (k+1) = \frac{k(k+1)}{2} + (k+1)$. By the distributive law this latter quantity is $(k+1)(\frac{k}{2}+1) = \frac{(k+1)(k+2)}{2}$.

Finally, then, the sum of the arithmetic progression is seen to be

$$n_0 + n_0 + d + n_0 + 2d + \ldots + n_0 + kd = (k+1)(n_0 + \frac{dk}{2}).$$

This formula was discovered as a schoolboy by Gauss.[2] Probably the way Gauss did this was by observing that the sum of the first and last terms is the same as the sum of the second and next to last term, ... So the only question is, how many terms are there and, in particular, is there a middle term?

[2]As busy work, the young Gauss was asked to compute the sum of numbers from 1 to 100 which, to the amazement of his teacher, he was able to do very quickly.

It might be remarked that another, more geometric, way to see that $\Sigma(k) = \frac{(k+1)(k)}{2}$ is the following. Let us imagine a square array of $k+1$ by $k+1$ dots. Clearly, the total number of dots is $(k+1)^2$, while the number on the diagonal is $k+1$. Thus, by symmetry, the number in either triangular area which results from removing the diagonal is $\frac{(k+1)^2-(k+1)}{2} = \frac{(k+1)(k)}{2}$. But the number of dots in either of these triangular areas is also clearly $\Sigma(k)$.

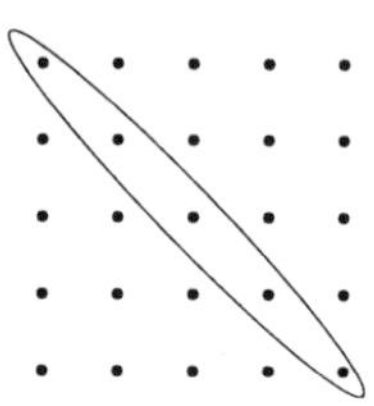

Figure 1.1

As an exercise it might be interesting to try and guess if there is a formula, and what it might be, for the sum if the first k squares $1^2 + 2^2 + \ldots k^2$. If your guess is correct you should be able to prove it by induction in a manner similar to the way we did for the sum of the first k integers. For $n \in \mathbb{Z}^+$, let $X_n = \{1, \ldots, n\}$. Define $P(X_n)$ to be the set of all subsets of X_n and S_n to be the set of all *permutations* of X_n, where by a permutation we mean a 1:1 and onto mapping or function from X_n to itself.[3] For small n try to calculate how many (distinct) subsets, or permutations, respectively there are. If you see a pattern, try to prove your guess by induction. Later we shall see how to deal with the number of subsets by means of the *binomial theorem*, which is itself proven by induction.

[3]If X and Y are sets, then a function $f : X \to Y$ is any rule which assigns to each $x \in X$ a definite $f(x) \in Y$. Such an f is called 1:1 and onto if each $y \in Y$ is $f(x)$ for a unique $x \in X$. A fuller discussion of these properties will be given in section 1.4

1.2 The Integers

Next we turn to the integers themselves. A feature of $\mathbb{Z}^+$ is that although we can add any two positive integers and still have a positive integer, this is not true of subtraction. Thus $4-9$ is not a positive integer. For this reason it is necessary to introduce the set $\mathbb{Z}$ of all integers.

$$\mathbb{Z} = \{\ldots -3, -2, -1, 0, 1, 2, 3, \ldots\}$$

Here these elements move both to the left and right. Clearly we can now subtract any integers and the result will be in $\mathbb{Z}$. Notice that although we still have an ordering (if $m \in \mathbb{Z}$ is to the right of $n \in \mathbb{Z}$ we say $n < m$), the principle of induction *no longer holds*. But there are a number of advantages here. For one thing there is an integer, namely 0, with the property that for all $n \in \mathbb{Z}$, $n + 0 = n$. Also we can now solve for $x \in \mathbb{Z}$ any equation of the form

$$a + x = b, \tag{1.2}$$

where a and b are given elements of $\mathbb{Z}$. Here is how: If $a + x = b$, then adding $-a$ (which exists in $\mathbb{Z}$, but not in $\mathbb{Z}^+$) to both sides yields $a+x-a = b-a$. Assuming that we can add the numbers on the left side in any order we get $x + a - a = x + 0 = x$. Thus if the original equation has any solution, then it could only be $b - a$. To test whether this is indeed a solution we have to substitute it into the original equation (1.2) and see if it satisfies. But, clearly $a + b - a$ is indeed b. A feature that persists in $\mathbb{Z}$ is the distributive law: $(a+b)c = ac+bc$. From this, taking $b = -a$, we see that $0 \cdot c = ac - ac = 0$. Thus for any $c \in \mathbb{Z}$, $0 \cdot c = 0$. We shall now take as a standing assumption that multiplication of integers a and b is *commutative*, i.e., $ab = ba$. The formalities of this situation is usually described by saying that the integers form a *commutative ring with identity*. Later we shall encounter many other interesting commutative rings with identity.

Another way of understanding the need to introduce zero and the negative integers is if we were interested in measuring some physical quantity, for example temperature. We would have to be prepared to deal with cold weather!

Exercise 1.2.1.

1. Using the distributive law show that $a^2 - b^2 = (a-b)(a+b)$ for any a and $b \in \mathbb{Z}$.

2. Try to guess and prove corresponding formulas for $a^n - b^n$ where n is any positive integer. Start with $n = 3$. If you are successful, then formulate and prove the general result.(There should be a telescoping sum with massive cancellation).

3. Suppose we call (a, b, c) a *Pythagorean triple* if a, b and $c \in \mathbb{Z}^+$ and $a^2 + b^2 = c^2$. This terminology is used because a, b and c form the sides and hypotenuse, respectively, of a *right triangle*. Show that taking $a = x^2 - y^2$, $b = 2xy$ and $c = x^2 + y^2$, where x and $y \in \mathbb{Z}^+$ and $x > y$ always gives a Pythagorean triple. Calculate a few of these for different values of x and y to convince yourselves that one gets all the familiar ones and at least one or two that may be unfamiliar.

We conclude this section with the binomial theorem. This is a formula expressing $(a+b)^n$, for any $n \in \mathbb{Z}^+$, as a "polynomial" in a and b with certain coefficients, which have some interest in their own right. Here we will take a and $b \in \mathbb{Z}$. But, as we shall see from the argument, this can be done much more generally as long as a and b commute.

Suppose r and $n \in \mathbb{Z}^+$ with $r \leq n$. We define $\binom{n}{r} = \frac{n!}{r!(n-r)!}$, where $s!$, which is called s *factorial*, means $s \cdot (s-1) \cdots 1$. For fixed, small n, it would be a good idea for the reader to calculate $\binom{n}{r}$ for all $0 \leq r \leq n$. We observe that $\binom{n}{r}$ is exactly the number of subsets consisting of r elements that can be chosen from $X_n = \{1, \ldots, n\}$. This is because there are n ways to fill the first spot. Having done so there are then $n-1$ ways to fill the second spot ... and finally, there are r ways to fill the last spot. Thus the number of possible sets of this type is exactly $n(n-1)(n-2)\cdots r$, if we distinguish the order of selection. But a permutation (see section on $\mathbb{Z}^+$) of a set doesn't change the set and as we saw there are exactly $r!$ such permutations. Therefore the number of possible sets of this type (without regard to their order) is exactly

$\frac{n\cdot(n-1)\cdot(n-2)\ldots r}{r!} = \binom{n}{r}$. In particular, $\binom{n}{r}$ is always an integer. Another consequence of this interpretation as the number of possible sets of r elements is that $\binom{n}{r} = \binom{n}{n-r}$. Why? Can you also prove this directly? The following simple lemma follows directly from the definitions and is also left as an exercise. From it we see that the $\binom{n}{r}$ can be generated in a very efficient and striking way by constructing *Pascal's triangle.*

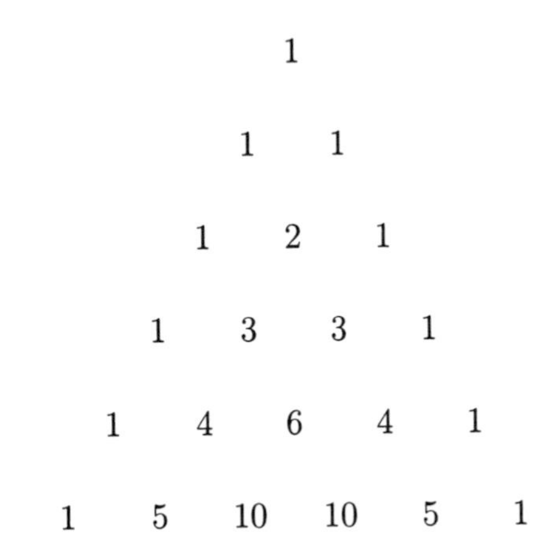

Figure 1.2: Pascal's triangle

Lemma 1.2.2. *For all r and $n \in \mathbb{Z}^+$ with $0 \leq r \leq n$, $\binom{n+1}{r} = \binom{n}{r} + \binom{n}{r-1}$.*

We now come to the binomial theorem itself. It plays an important role in *probability*. Later, when we look at the geometric series, we shall see analogues of this result for negative integer values of the exponent.

Theorem 1.2.3. *For each $n \in \mathbb{Z}^+$ and any a and b which commute we have*

$$(a+b)^n = \sum_{r=0}^{n} \binom{n}{r} a^r b^{n-r} \tag{1.3}$$

Proof. The proof will be by induction on n. Clearly (1.3) holds for $n = 1$. Assuming it holds for n, multiply (1.3) by $a + b$ and get

$$(a+b)^{n+1} = \sum_{r=0}^{n} \binom{n}{r} a^{r+1} b^{n-r} + \sum_{r=0}^{n} \binom{n}{r} a^r b^{n+1-r}.$$

Letting $s = r + 1$ in the first summand gives $\sum_{s=1}^{n+1} \binom{n}{s-1} a^s b^{n+1-s}$. Now renaming s as r, adding this last term to the second summand and making use of the lemma proves (1.3) for $n + 1$. □

Corollary 1.2.4. *If $b \geq 0$ then $(1+b)^n \geq 1 + nb$ for all $n \geq 1$.*

Proof. In the above let $a = 1$. □

As an exercise try to formulate a *multinomial* theorem. Then, relying on the binomial theorem, prove this by induction on the number of variables. The following gives an alternative proof of our result concerning the number of subsets of X_n.

Corollary 1.2.5. *The number of subsets of X_n is 2^n.*

Proof. As we saw above, $\binom{n}{r}$ is the number of subsets of r elements of X_n. Hence $\sum_{r=0}^{n} \binom{n}{r}$ is the total number of subsets of X_n. Taking $a = 1 = b$ in the binomial theorem we see that this is just 2^n. □

Exercise 1.2.6. *Prove that:*

(i) $\frac{(2k+1)!}{2^k k!} = (2k+1)(2k-1)\cdots 1$.

(ii) $2^k k! = (2k)(2k-2)\cdots 2$.

Observe that multiplying these two gives a tautology, that is a statement which is self evidently true. Therefore it is really only necessary to prove one of these statements.

1.3 The Rational Numbers

Unfortunately, all this is not enough because we cannot always divide integers and have the quotient an integer. For example $\frac{2}{3}$ is not an integer. For this reason it will be necessary to introduce the set $\mathbb{Q}$ of rational numbers. These are, by definition, things of the form $\frac{b}{a}$ where a and b are integers and $a \neq 0$. We shall see shortly why this latter restriction is unavoidable. We must also realize that $\frac{b}{a}$ could equal $\frac{d}{c}$, namely exactly when $bc = ad$.

In particular, if $c \neq 0$ then

$$\frac{b}{a} = \frac{bc}{ac},$$

i.e., we can cancel out the c. This is because $abc = bac$. For example $\frac{2}{3} = \frac{4}{6}$. We understand $\mathbb{Z}$ to be those things of the form $\frac{b}{1}$. From this we see that $1 \cdot c = c$ for each $c \in \mathbb{Q}$.

In $\mathbb{Q}$ we add and multiply as follows: $\frac{b}{a} + \frac{d}{c} = \frac{bc+ad}{ac}$, $\frac{b}{a} \cdot \frac{d}{c} = \frac{bd}{ac}$. We shall see in a moment that since a and $c \neq 0$ then also $ac \neq 0$. Now the distributive law also holds in $\mathbb{Q}$. That is $\frac{f}{e}(\frac{b}{a} + \frac{d}{c}) = \frac{f}{e} \cdot \frac{b}{a} + \frac{f}{e} \cdot \frac{d}{c}$. To see that this is so apply the definitions of addition and multiplication given just above, together with the distributive law in $\mathbb{Z}$.

Since the proof of the fact that $0 \cdot c = 0$ for $c \in \mathbb{Z}$ only depended on the distributive law, which we see now also holds in $\mathbb{Q}$, it follows that $0 \cdot c = 0$ for $c \in \mathbb{Q}$. We shall take $\frac{b}{a} < \frac{d}{c}$ to mean that $bc < ad$ (in $\mathbb{Z}$). Thus $\frac{2}{3} < \frac{3}{4}$ since $2 \cdot 4 < 3 \cdot 3$. It is also worth mentioning, and the reader should check this, that the ordering on $\mathbb{Z}$ is consistent with this one on $\mathbb{Q}$.

Now an important feature in $\mathbb{Q}$, analogous to finding negatives in $\mathbb{Z}$, but which is not present in $\mathbb{Z}$, is the existence of inverses of non zero elements. $\frac{b}{a} \cdot \frac{a}{b} = \frac{ba}{ab} = 1$. Thus each nonzero element of $a \in \mathbb{Q}$ has a multiplicative inverse $b \in \mathbb{Q}$ such that $ab = 1$ and we write a^{-1}. Notice in contrast that, in $\mathbb{Z}$ the number 2 has no multiplicative inverse. In particular, as remarked above, if $ac = 0$ where a and c are $\in \mathbb{Z}$, or indeed $\in \mathbb{Q}$, then one or the other of these (or both) must be zero. For suppose $a \neq 0$; then we simply multiply $ac = 0$ by a^{-1} and get $a^{-1}ac = a^{-1}0 = 0$ while the left side is $a^{-1}ac = 1c = c$. Thus if $a \neq 0$, then $c = 0$.

We now come to the main reason for introducing the rational numbers. Given a and $b \in \mathbb{Z}$, we would like to, but can't always, solve the equation $ax = b$ for an $x \in \mathbb{Q}$. Now when can we expect to be able to solve such an equation? If $a = 0$ and $b \neq 0$, then for any $x \in \mathbb{Q}$, $ax = 0 \neq b$. Thus we must assume that $a \neq 0$ or we have no chance. The question is can we solve this equation when $a \neq 0$? The answer is yes; since $a \neq 0$ we can multiply $ax = b$ by a^{-1} and get $a^{-1}ax = 1x = a^{-1}b$. Thus if there is any solution there is only one and it must be $x = \frac{b}{a}$. Substituting as above we see that this is indeed a solution. Notice this solution may not lie in $\mathbb{Z}$.

Another way of understanding the need to introduce fractions is if we were interested in measuring some physical quantity, for example the length of a board. A carpenter's tape cannot only have integers (say in inches) on it since the length to be measured may fall in between two successive integers. Usually such tapes measure to the nearest sixteenth of an inch. Of course the length of a particular board may also fall between successive sixteenths or indeed of any other unit which would have to be specified in advance. In fact, as we shall see, it may not fall on a rational mark at all! We will return to this question when we discuss the real numbers in the next section. Just as with the integers, we will think of the rational numbers as arranged on the number line and of course taking up much more of it. In fact we shall see later that any point on the line can be approximated to any desired degree of accuracy by rational numbers. It might be mentioned that thinking of numbers as points on a line and vice-versa is an important idea of René Descartes, the consequences and generalizations of which have not yet been fully played out. We shall now summarize the properties of $\mathbb{Q}$ which we have

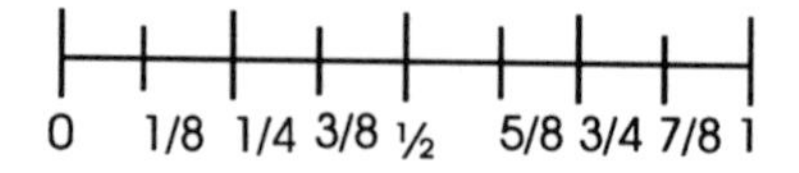

Figure 1.3

discussed. $\mathbb{Q}$ has two operations, addition and multiplication. They satisfy. $a + b = b + a$, $a + (b + c) = (a + b) + c$, $ab = ba$, $a(bc) = (ab)c$, and are connected by the distributive law $a(b+c) = ab+ac$. Each $a \in \mathbb{Q}$ has an additive inverse $-a$ such that $a + (-a) = 0$ and each $a \neq 0 \in \mathbb{Q}$

has a multiplicative inverse a^{-1} such that $aa^{-1} = 1$. Something that has these properties is called a *field*. In a field F there exist unique solutions to the equations $a + x = b$, where a and $b \in F$ and

$$ax = b, \tag{1.4}$$

where $a \neq 0$ and $a, b \in F$. We shall have occasion to take a look at some other fields later.

Notice that in a field F, $(-a)(b) = -ab$ and $(-a)(-b) = ab$, where a and $b \in F$. Since $ab + (-a)(b) = (a + (-a))b = 0b = 0$, we see by uniqueness of the solution to (1.2) that $(-a)(b) = -ab$. Similarly $(-a)(-b) + (-a)(b) = 0$ and $(-a)(b) + ab = 0$. Subtracting yields $(-a)(-b) - ab = 0$ and as above we get $(-a)(-b) = ab$. Finally, we observe that in a field $1 \neq 0$. For if $1 = 0$, then for any $a \in F$, $a \cdot 1 = a \cdot 0$, so $a = 0$. Thus we would only have a single element, namely 0. We shall understand fields to have at least two elements.

Exercise 1.3.1. An interesting application of solving equations such as (1.4) is the following problem: Find all the times on a clock when the minute and hour hand assume the same position. Of course, 12 noon is one such time. But shortly after 1:05 is another, etc. Hint: The rate at which the hour hand turns is $\frac{1}{12}$ revolution/hour while that of the minute hand is 1 revolution/hour. Starting at 1 o'clock, after time t the position of the hour hand will be $\frac{1}{12} + \frac{1}{12}t$. After time t the position of the minute hand will be $1t$. They will therefore coincide when $t = \frac{1}{11}$ hour or $\frac{60}{11}$ min. Thus, the first time the hands coincide is at $1:054545\ldots$.

Over a field we can also solve systems of linear equations, not merely individual ones. For example, we can consider a system of two equations in two unknowns x and y as follows.

$$ax + by = e$$

$$cx + dy = f$$

where $a, b, c, d, e, f \in F$. We multiply the first equation by c and the second one by a, and then subtract getting $(ad-bc)y = af-ce$. Hence, as above, if $ad-bc \neq 0$, we can solve this last equation and get $y = \frac{af-ce}{ad-bc}$. Similarly, eliminating y yields $x = \frac{dc-bf}{ad-bc}$. Thus if $ad-bc \neq 0$, the system has a unique solution. As an exercise show by an example that if $ad-bc=0$, the system of equations either has no solution or infinitely many solutions. Later we shall see what the geometric significance is of the existence or non existence of solutions. More or less the same reasoning and geometric significance applies to systems of n equations in n unknowns where $n > 2$. Notice that the field here is arbitrary, i.e. need not to be $\mathbb{Q}$(see the last paragraph of section 1.8).

Exercise 1.3.2. If F is the temperature in Fahrenheit and C in Celsius then

$$F = \frac{9}{5}C + 32.$$

Find the temperature when $F = C$.

1.4 The Real Numbers

If $p(x) = a_n x^n + a_{n-1}x^{n-1} + \ldots + a_0$ where $a_n \neq 0$, we say p is a *polynomial*. The a_i are called its *coefficients* and n is its *degree*, written $\deg p$. We next consider some *polynomial equations*, of degree higher than 1. For example

$$x^2 - 2 = 0. \tag{1.5}$$

(1.5) is an equation with coefficients in $\mathbb{Q}$ (in fact in $\mathbb{Z}$) and the degree here is 2. If a positive number x has the property that $x^2 = 2$ we shall say that x is the square root of 2 and write $x = \sqrt{2}$. Clearly, $x = \sqrt{2}$ is a solution to (1.5). But since as we saw above $(-\sqrt{2})(-\sqrt{2})$ also equals 2, this equation has at least two solutions. We shall see later that a polynomial equation of degree 2 has at most two solutions; hence in this case exactly two solutions. In this sense there is an important difference between polynomial equations of degree 1 and those of higher degree. Now recalling the Pythagorean Theorem (we will prove this in chapter 3) we see that $\sqrt{2}$ is on the number line somewhere between

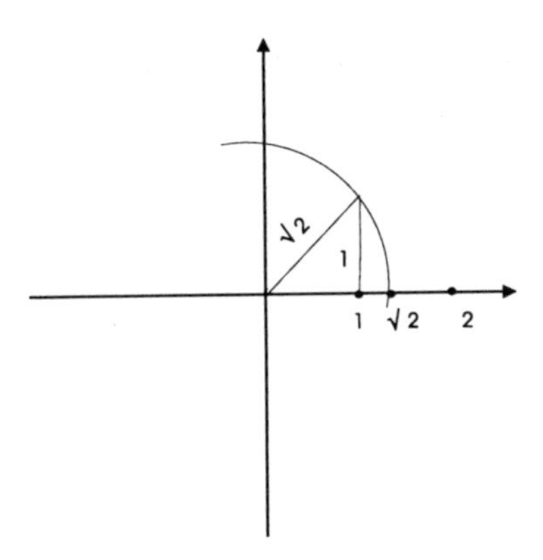

Figure 1.4

1 and 2(see figure 1.4) . We will now show that nevertheless it is not $\in \mathbb{Q}$, thus proving that there are *irrational numbers*. Suppose on the contrary that $\sqrt{2} = \frac{a}{b}$, where a and $b \in \mathbb{Z}$. By cancelling them out we may clearly assume that a and b have no common factors. Squaring both sides gives $2 = \frac{a^2}{b^2}$ and multiplying by b^2 yields $2b^2 = a^2$. Thus the integer a has the property that a^2 is an integer multiple of 2.

Lemma 1.4.1. *Suppose the integer n has the property that $n^2 = 2k$ for some integer k. Then n is itself a multiple of 2.*

Proof. If not, then $n = 2j + 1$ for some integer j (n is odd). Squaring, we get $n^2 = 4j^2 + 4j + 1 = 2(2j^2 + 2j) + 1$, an odd integer. On the other hand this is $2k$, an even integer. This contradiction shows that our original assumption must be wrong and n is indeed even. □

Returning to the proof of the irrationality of $\sqrt{2}$ we now see that since a^2 is a multiple of 2 so is a. Thus $a = 2j$, where j is an integer. Squaring we get $a^2 = 4j^2 = 2b^2$, so $b^2 = 2j^2$. But then, applying the lemma again tells us that b is also even. This means that both a and b have 2 as a common factor, a contradiction.

Exercise 1.4.2. Suppose a craftsman is tiling a floor with square tiles. He wishes to make a decorative border by diagonally laying these same tiles along the edge. He starts by having the corners line up. Can the corners ever line up again?

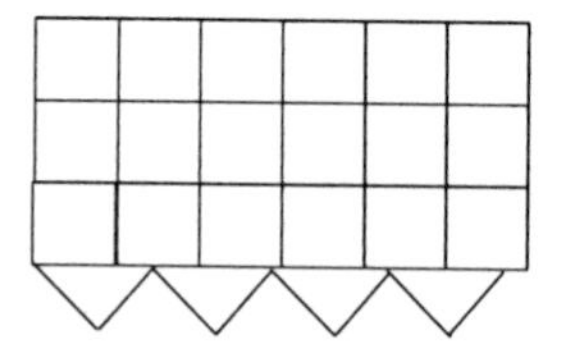

Figure 1.5

The existence of irrational numbers was a discovery of the ancient Greeks which they found to be very upsetting. This was because they supposed that "geometry" (points on a line) was "in harmony" with "arithmetic"(rational numbers). Perhaps this was because, as we shall soon see, you can approximate any point on the number line by a rational number, you just can't always hit it!

Filling in all the holes in the number line occupied by the *many* irrational numbers gives us $\mathbb{R}$, the *real* number field. Over $\mathbb{R}$ we can now find solutions to many (but not all) polynomial equations. For example, suppose we want to find a solution or *root* of a polynomial equation of odd degree, say $p(x) = 0$, with coefficients in $\mathbb{R}$. Since by assumption the leading coefficient is different from zero and as we saw earlier dividing by this constant doesn't affect the roots, we may assume that the leading coefficient is 1. Thus

$$p(x) = x^n + a_1x^{n-1} + a_2x^{n-2} + \cdots + a_n = 0$$

where n is odd. Now dividing by x^n we get

$$\frac{p(x)}{x^n} = 1 + \frac{a_1}{x} + \frac{a_2}{x^2} + \cdots + \frac{a_n}{x^n} = 0$$

We shall see later that as x runs through larger and larger real numbers (we say $x \to +\infty$), each of the terms of the polynomial after the first, and therefore also their sum, becomes as small as one likes. So

$\lim_{x\to+\infty} \frac{p(x)}{x^n} = 1$. Similarly $\lim_{x\to-\infty} \frac{p(x)}{x^n} = 1$. This means that as $x \to \pm\infty$, $p(x)$ and x^n look roughly the same. Now just what does $q(x) = x^n$ look like when n is odd ? Since we have a continuous function [4] $p(x)$ which must be positive for a certain value of x and negative for some other value of x, it must have a zero. In fact, all intermediate values must be assumed. This is called the *principle of continuity*. From it we see that $p(x)$ must have a zero.

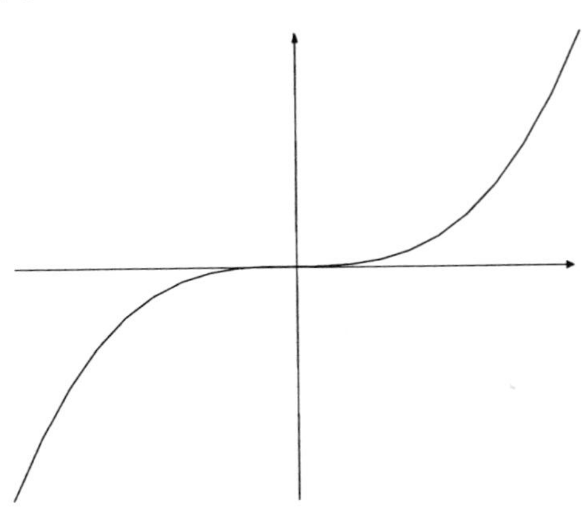

Figure 1.6: $y = x^3$

Exercise 1.4.3.

1. Is the statement above true when the degree of p is even, e.g. 2, or does something go wrong?

2. Suppose p and q are polynomials of the same degree, say n, with leading coefficients a_n and b_n, respectively. Show $\lim_{x\to\infty} \frac{p(x)}{q(x)} = \frac{a_n}{b_n}$.

Let X and Y be two sets. One says that $f : X \to Y$ is a *mapping* or *function* between these sets if to each $x \in X$ corresponds a definite

[4] We do not give a precise definition of continuous function here. It will be sufficient for us to know that polynomials are continuous and later that the cosine function is continuous.

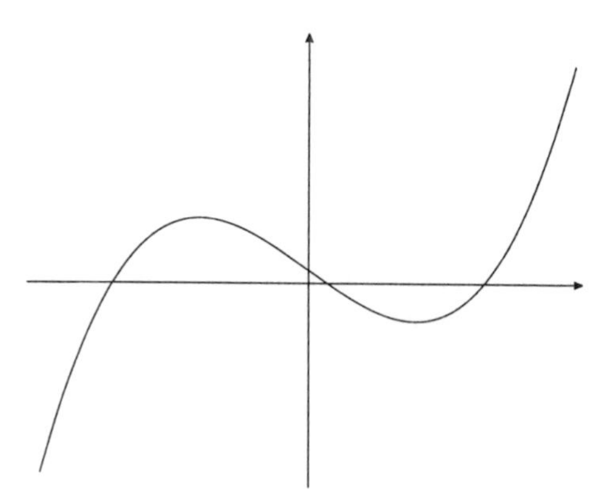

Figure 1.7: a polynomial of degree 3

element of Y called $f(x)$. Thus the mapping is rule that makes this assignment. If every element of Y is $f(x)$ for some $x \in X$ then f is said to be *onto* or *surjective*. $f : X \to Y$ is said to be *1:1* or *injective* if two distinct elements(or points) of X can not be sent by f to the same element (or point) of Y.

For example the map $f(x) = x^2$, as a map from $\mathbb{R}$ ro $\mathbb{R}$ is not onto since it leaves out all the negative numbers, whereas as map from $\mathbb{R}$ to $\mathbb{R}^+$ (the non-negative reals) it is onto. Moreover, in neither case is this map 1:1 because the square of a negative number is the same as the square of the corresponding positive number. On the other hand, the *linear functions* of the form $f(x) = ax + b$, where $a \neq 0$, are 1:1 and onto as a map from $\mathbb{R}$ to itself. Similarly, the map $f(x) = x^3$ as a map from $\mathbb{R}$ to itself is also both 1:1 and onto.

If X is a finite set and f a map from X to itself, then f is 1:1 if and only if it is onto.

The following is an interesting fact connected with the ideas immediately above.

Theorem 1.4.4. *Suppose $p(x)$ is a polynomial with real coefficients and as a map $\mathbb{R} \to \mathbb{R}$, p is 1:1. Then p is onto.*

Proof. First we can see that the degree of p must be odd. This is because if the degree were even then, by an argument above, for large values of x and $-x$ the function takes on all negative or all positive values depending on the sign of the leading coefficient. Therefore, in any case, p can not be 1:1. For definiteness we assume the leading coefficient is positive. Since the degree of p is odd as x gets large, $p(x)$ goes to ∞ and as x gets smaller and smaller, $p(x)$ goes to $-\infty$. Therefore, by the reasoning above, everything beyond those values of $p(x)$ gets hit. But the things between those values also get hit by the principle of continuity and so p is onto. $\square$

We remark that in the case of polynomial in n variable this an unsolved problem.

We now come to another important principle of continuity of the real numbers, usually called the *squeezing principle*. Namely, suppose we have two other functions $g(x)$ and $h(x)$ in addition to a function $f(x)$ such that for all x near a, $g(x) \leq f(x) \leq h(x)$. If $g(x)$ gets nearer and nearer to some real number b as $x \to a$ and the same is true of $h(x)$, then $f(x)$ also must get nearer and nearer to b as $x \to a$. This is because $f(x)$ is crushed in between $g(x)$ and $h(x)$ and so has nowhere else to go.

A further example of filling in all the holes as well as what happens to a quantity, say $f(x)$, as the real number x gets larger and larger, or smaller and smaller, or, more generally, nearer and nearer to, say, a (in this case we write $x \to a$), is the fact that $\frac{\sin x}{x} \to 1$ as $x \to 0$. This latter result is useful in calculus.

To see this is so let $f(x) = \frac{\sin x}{x}$, where $x \neq 0$. For a fixed angle x consider the following diagram , where here we assume $x > 0$. (A similar diagram would be drawn if $x < 0$). Each of the regions in the diagram is contained in the next one so their areas are related by $A_1 \leq A_2 \leq A_3$. Using the obvious fact that the area of a triangle is one half the base times the height and that the area of a circle of radius r is πr^2, while that of a sector of a circle is proportional to its opening angle, it follows that $\frac{1}{2}\sin x \leq \pi 1^2 \frac{x}{2\pi} \leq \frac{1}{2}\tan x$. Since $x > 0$, we know $\sin x$ is also positive and multiplying by $\frac{2}{\sin x}$ gives $1 \leq \frac{x}{\sin x} \leq \frac{1}{\cos x}$. So

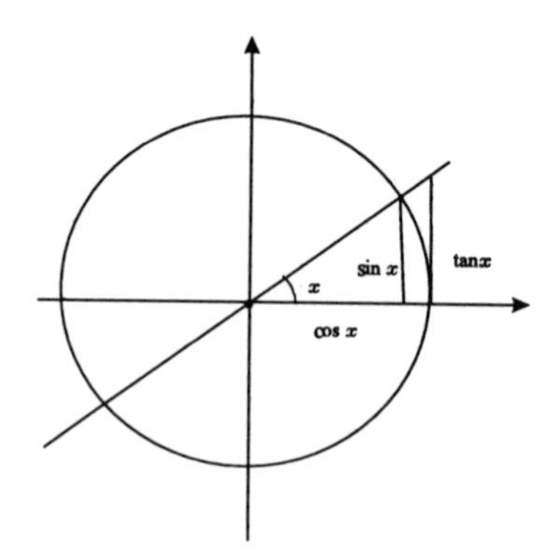

Figure 1.8

taking reciprocals we get

$$\cos x \leq \frac{\sin x}{x} \leq 1.$$

Since $\cos x = \cos(-x)$ and $\frac{\sin x}{x} = \frac{\sin(-x)}{-x}$, this also holds if $x < 0$. To follow, in detail, the manipulations involving inequalities used here, see the properties of an ordered field given on the page below. Now $\cos x \to 1$ as $x \to 0$ so the squeezing principle tells us the same is true of $\frac{\sin x}{x}$. Notice that the fact that $\lim_{x\to\infty} \frac{\sin x}{x} = 1$ itself implies that the area of a circle of radius r is πr^2. This is done by a process known to the ancient Greeks as *the Method of Exhaustion*. It works as follows:

$$A_n = r^2 n \cos\frac{\pi}{n} \sin\frac{\pi}{n} = r^2 n \frac{\sin\frac{2\pi}{n}}{2}$$

$$A = \lim_{n\to\infty} A_n$$

Let $\frac{2\pi}{n} = x$. Then $n = \frac{2\pi}{x}$.

$$A = \lim_{x\to 0} \frac{r^2 2\pi}{2} \frac{\sin x}{x} = \pi r^2 \lim_{x\to 0} \frac{\sin x}{x} = \pi r^2.$$

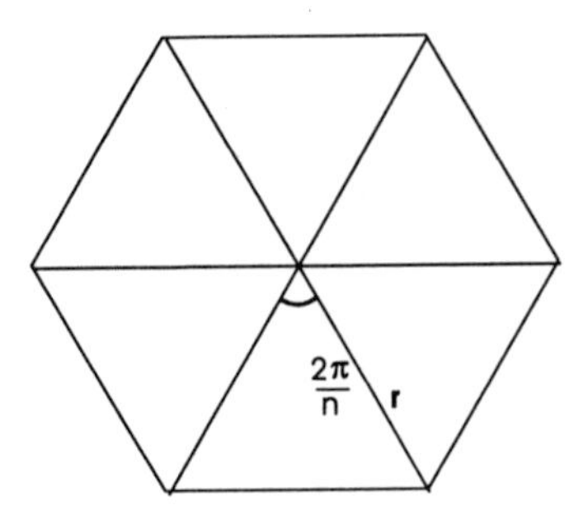

Figure 1.9

For $a \neq 0 \in \mathbb{R}$ and $n \in \mathbb{Z}^+$, we define a^n to mean a multiplied by itself n times. Also, $a^0 = 1$ and $a^{-n} = (a^{-1})^n$. This defines a^n for all $n \in \mathbb{Z}$. Evidently, for all $n \in \mathbb{Z}$, $a^{n+m} = a^n a^m$ and $(a^n)^m = a^{nm}$. Prove these *laws of exponents* as an exercise. We can extend the definitions of exponentiation from $\mathbb{Z}$ to $\mathbb{Q}$ as follows: If $r = \frac{n}{m} \in \mathbb{Q}$, and $a > 0$, then a^r is to mean $(a^n)^{\frac{1}{m}}$, where a fractional power means take the appropriate root. We shall see later that we must assume $a > 0$ because although we can't take even roots of negative real numbers, although we can take any root of a positive real number. As an exercise show that for $a > 0 \in \mathbb{R}$ and $q, r \in \mathbb{Q}$, $a^{q+r} = a^q a^r$ and $(a^q)^r = a^{qr}$.

Notice that the ordering on $\mathbb{Q}$ extends in a natural way to $\mathbb{R}$; that is, just as in $\mathbb{Q}$, $a < b$ means that a lies to the left of b on the number line. It might now be worthwhile to point out some of the arithmetic properties of this ordering. A field satisfying the following *axioms* is called an *ordered* field.

1. For any $a \in \mathbb{R}$, either $a < 0$, $a = 0$ or $a > 0$.

2. If $a \leq b$ and $b \leq c$ then $a \leq c$.

3. If $a < b$, then $a + c < b + c$ for any real c.

4. If $a < b$, then $ac < bc$ for any real $c > 0$.

For a real number a, or indeed an element in any ordered field, we now define $|a|$ to be a, if $a \geq 0$ and $-a$ otherwise. Then we have the following properties which the reader is asked to verify as an exercise. In this connection observe that for any $x \in \mathbb{R}$, $|x| \geq x$, $|x|^2 = x^2$ and that if x and $y \geq 0$, then $x \leq y$ if and only if $x^2 \leq y^2$.

1. $|a| \geq 0$.

2. $|a| = 0$ only if $a = 0$.

3. For all $a, b \in \mathbb{R}$, $|ab| = |a| \cdot |b|$.

4. For all $a, b \in \mathbb{R}$, $|a + b| \leq |a| + |b|$. (This last item is called the *triangle inequality*).

We can now explain why it is that for any $a \in \mathbb{R}$, $\lim_{x \to +\infty} \frac{a}{x^n} = 0$. Using 3) we see that $|\frac{a}{x^n}| = \frac{|a|}{|x|^n}$. But if we take $x \geq 1$, we saw before that $x^n \geq x$ and hence $\frac{|a|}{|x|^n} \leq \frac{|a|}{|x|}$. Since a is fixed, as $x \to +\infty$ this latter term becomes as small as we want.

As a further application of the principal of continuity we show that for any $n \in \mathbb{Z}^+$, the positive real number a has a real n*th* root in $\mathbb{R}$. We consider the polynomial $p(x) = x^n - a$, where $n \geq 2$ and the equation $p(x) = 0$. Clearly $p(0) = -a < 0$. For very large positive x we see that $p(x) = x^n - a \to \infty$, and therefore is as large as one likes. In particular it is positive. By the principle of continuity it must have a zero.

We now study elementary properties of an ordered field. First of all, in an ordered field $0 < 1$. We have already seen that $0 \neq 1$. The only other possibility is $1 < 0$. But then, by 3), $0 < -1$. Hence by 4) $-1 < 0$. This contradicts 1), proving $0 < 1$. We show if $a > 0$, then $a^{-1} > 0$. Since $aa^{-1} = 1$, it follows that $a^{-1} \neq 0$. Suppose $a^{-1} < 0$. Multiplying by $a > 0$ would then give $1 < 0$, a contradiction. From this it follows easily that if $0 < a \leq b$, then $b^{-1} \leq a^{-1}$. (Just multiply both sides of the given inequality by the positive number ab.) Finally, suppose $b \geq 1$ and $n \in \mathbb{Z}^+$. Then $b^n \geq b$ and therefore $\frac{1}{b^n} \leq \frac{1}{b}$. Multiplying $b \geq 1$ by the positive number b gives $b^2 \geq b$. Repeating this gives $b^3 \geq b^2$ so by induction for all n, $b^n \geq b^{n-1}$. Hence by 2), $b^n \geq b$.

Given two positive real numbers x and y we call $\frac{x+y}{2}$ their *arithmetic mean* and $\sqrt{xy}$ their *geometric mean*. As an exercise show $\sqrt{xy} \leq \frac{x+y}{2}$ and that equality holds if and only if $x = y$. Can you think of a generalization of this to more than two numbers?

From the Euclidean algorithm (to be proved in Section 2.5) it also follows that

Corollary 1.4.5. *If a and $b > 0$, then there is some $n \in \mathbb{Z}$ such that $na > b$.*

(In other words, no matter how small $a > 0$ is, nor how big b is, some multiple of a is bigger than b). This is called the *Archimedian* property. It holds because otherwise for all $n \in \mathbb{Z}$ we would have $0 \leq na \leq b$. Since by the above $a^{-1} > 0$, multiplying we would get $0 \leq n \leq \frac{b}{a}$, a contradiction since the positive integers move infinitely to the right.

Corollary 1.4.6. *Any real number can be approximated to any desired degree of accuracy by rational numbers. Or, to put it another way, between any two real numbers $x < y$,* no matter how close, *there is always a rational number.*

Proof. Since $x < y$, we know $y - x > 0$ and so the Archimedean property tells us we can find a positive integer n such that $n(y - x) > 1$. But because $ny - nx > 1$, there must be an integer m in between: $ny > m > nx$. Multiplying this inequality by the positive number $\frac{1}{n}$ tells us $y > \frac{m}{n} > x$. □

Finally, we define the meaning of the *decimal expansion of a real number*. Clearly, any real number x falls between two successive integers, say n and $n + 1$. Since n is uniquely determined by x we call it the *integer part* of x. Dividing the interval from n to $n+1$ into 10 equal parts and successively labelling them $0, 1, 2, \ldots, 8, 9$, we see in a similar manner to the above that x lies in exactly one of these intervals, with the understanding that if it lies on the boundary we count this in favor of the right side. This gives us the first decimal place of x. Continuing to divide this interval into 10 equal parts and labelling them $0, 1, 2, \ldots, 8, 9$ as before and, using the same favoring the right convention, we get the second decimal place of x. Continuing in this manner indefinitely gives the decimal expansion of

$$x = n.a_1a_2 \cdots a_k \cdots.$$

Writing this last equation indicates implicitly that not only does x give rise to its decimal expansion, but that conversely a decimal expansion determines a unique real number x. This last statement is true because we keep dividing the interval into smaller and smaller pieces and the decimal expansion constitutes a set of better and better instructions as to where x is located. But we are not yet in a position to actually prove this; and in fact this last statement can be viewed as one of the various alternative (formal) definitions of the real numbers.

Definition 1.4.7. A set is called countable or enumerable if it is in 1:1 correspondence with the set of integers.

We now discuss the famous theorem of Georg Cantor that the real numbers, $\mathbb{R}$ cannot be enumerated . That is, cannot be put in 1:1 correspondence with the positive integers, $\mathbb{Z}^+$. To do this it is clearly sufficient to show that the unit interval, $[0, 1]$, itself cannot be enumerated. Just as Cantor did, this will is done by use of his celebrated *diagonal process* .

Theorem 1.4.8. $\mathbb{R}$ *cannot be enumerated.*

Proof. This will be proved by contradiction. Suppose $[0, 1]$ can be enumerated. Let each real number between 0 and 1 be represented by a decimal. In the event that the decimal expansion is not unique (such as $1.0000\ldots = .9999\ldots$) we select the terminating one. Let

$$.x_{n,1}x_{n,2}x_{n,3} \ldots x_{n,j} \ldots$$

be the n*th* one. Chose a number between 0 and 9 which is different from $x_{1,1}$ and from 9 to be the first decimal of a real number in $[0, 1]$ we are in the process of constructing. Similarly, chose a number between 0 and 9 which is different from $x_{2,2}$ and from 9 to be the second decimal. Continuing in this way we get something in $[0, 1]$ which cannot be any of the enumerated numbers since for each integer n it differs from the n*th* one in the n*th* spot. This contradiction proves the theorem. Of course since $[0,1]$ is not enumerable neither is $\mathbb{R}$. □

1.5 The Integers Revisited

We now use the Archimedean property (of $\mathbb{R}$) to derive some facts about $\mathbb{Z}$ itself. First we turn to what is known as the *Euclidean algorithm* or *division algorithm.* The r below is called the *remainder*.

Theorem 1.5.1. *Given $a, b \in \mathbb{Z}$ with $a > 0$ there is always some $n, r \in \mathbb{Z}$ with $0 \leq r < a$ such that $b = na + r$.*

To prove the Euclidean algorithm let $S = \{n \in \mathbb{Z} : an > b\}$. By the above $S \neq \phi$. By the Principle of Induction, let n_0 be its smallest element. Then $a(n_0 - 1) \leq b < an_0$. Let $r = b - a(n_0 - 1)$. Then $b = na + r$ and $0 \leq r$. If $r \geq a$, then $b - a(n_0 - 1) \geq a$; so we would get $b \geq an_0$, a contradiction.

Exercise 1.5.2. Show that for a fixed a and b, the n and r in the Euclidean algorithm are unique.

We now define a *prime* number. This is a positive integer, p, greater than 1 which can't be factored into a product of two or more numbers in $\mathbb{Z}^+$ (except for the trivial factorization $p = p \cdot 1$).

Theorem 1.5.3. *If a prime p divides ab, where a and $b \in \mathbb{Z}$, then p divides a or b (or both).*

Notice that it is essential here that p be a prime. For example, 6 divides $3 \cdot 2$, but 6 divides neither 3 nor 2.

Proof. If p doesn't divide a we will show that it must divide b. Let $S = \{np + ma : n, m \in \mathbb{Z}\}$ and d be the smallest positive integer in S. Then $d = n_0p + m_0a$. By the Euclidean algorithm write $p = kd + r$ where $k, r \in \mathbb{Z}^+$ and $0 \leq r < d$. Then $r = p - k(n_0p + m_0a) = (1 - kn_0)p + m_0a$, that is $r \in S$. Since $r < d$ and d is the smallest such positive integer, we conclude that $r = 0$ and therefore $p = kd$. But since p is a prime, either $d = p$ or $d = 1$.

Suppose $d = p$. Applying the Euclidean algorithm to a and d tells us $a = jd + r_1$ where $0 \leq r_1 < d$. This gives us something in S, namely r_1, smaller than d. Therefore $0 = r_1$ and $a = jp$, a contradiction since we are assuming that p doesn't divide a. Therefore we must have $d = 1$; that is $n_0 p + m_0 a = 1$. Multiplying by b yields $n_0 pb + m_0 ab = b$. Since p divides ab and it evidently also divides $n_0 pb$, it must divide b. □

Corollary 1.5.4. *(The fundamental theorem of arithmetic). Any integer in $\mathbb{Z}^+$ can be factored uniquely into primes.*

Proof of existence. Suppose there were some $n \in \mathbb{Z}^+$ which couldn't be factored into primes. Let n_0 be the least such integer. In particular, n_0 can't be a prime. Therefore it can be factored $n_0 = ab$. This means that since neither a nor $b = 1$, both a and $b < n_0$. Thus they can each be factored into primes. Putting these two factorizations together shows that n_0 can be factored into primes.

Proof of uniqueness. Let $n = p_1 p_2 \cdots p_r$ be a factorization of the integer n into primes. Suppose there were another such factorization $n = q_1 q_2 \cdots q_s$. Since p_1 divides n, it divides q_i for some $i = 1, \ldots, s$. By reordering the index set $i = 1, \ldots, s$ we may assume that p_1 divides q_1. But then $p_1 = q_i$. Multiplying by the inverses (in $\mathbb{Q}$) we reduce ourselves to a problem with index set of lower order. Just keep going by induction and, after reordering, get $r = s$ and $p_i = q_i$ for all i.

In view of the uniqueness of the prime factorization of an integer n we can group the prime divisors and write $n = p_1^{e_1} \cdots p_k^{e_k}$, the e_i being positive integers. Each e_i is called the *multiplicity* with which the prime p_i occurs.

Corollary 1.5.5. *There are infinitely many primes.*

Proof. Suppose, to the contrary that there are only finitely many primes, $\{p_1, \ldots, p_k\}$. Then since the integer $n = p_1 p_2 \ldots p_k + 1$ must factor into primes, one of the p_i must divide n. Now each p_j divides $p_1 p_2 \ldots p_k$ so p_i divides 1. This is impossible. Hence there are infinitely many primes. □

This last result has a generalization due to the 19th-century mathematician L. Dirichlet which states that there are infinitely many primes in any arithmetic progression in $\mathbb{Z}$ with leading term n_0 and difference d, so long as n_0 and d are relatively prime. This requirement is necessary; the even integers, for example, contain only the prime 2.

For integers a and b, $a \neq 0$, we shall say $a|b$ (to be read a divides b), if b is an integer multiple of a.

Definition 1.5.6. The greatest common divisor of $a_1, a_2....a_n$, where $a_i \in \mathbb{Z}$, denoted by $gcd(a_1, a_2, ...a_n)$ is the positive integer number d with the following properties:

(i) $d|a_i$ for every i.

(ii) For any d' that satisfies condition (i) then we have $d'|d$.

As $d'|d$ implies that $d' \leq d$ therefore d is the greatest among the common divisors of a_i's, and automatically unique. Since the set of positive divisors is finite then there is always such a maximum. This proves the existence of the greatest common divisor.

Definition 1.5.7. Two integer numbers a and b are said to be relatively prime if $gcd(a, b) = 1$.

Theorem 1.5.8. *Let a and b be two relatively prime integers. Then there are integers r and s such that $ra + sb = 1$.*

Proof. Consider the set $S = \{xa + yb : x, y \in \mathbb{Z}\}$. S is non empty as $\pm x$ $\pm y$ are in S.
Let d be the smallest positive number in S. Let d' be an arbitrary element of S then using division algorithm there are p and q such that

$$d' = pd + q$$

where $0 \leq q < d$.
Since d and d' are in S it follows that q is also in S. But since d is lowest positive integer in S it follows that $q = 0$ or in other words $d|d'$. In particular $d|x$ and $d|y$. By assumption x and y are relatively prime, therefore $d = 1$. This proves the assertion. □

Corollary 1.5.9. *Let a and b be two integers and $d = gcd(a,b)$. Then there are integers r and s such that*

$$d = ra + sb.$$

Proof. Since $d = gcd(a,b)$ then $\frac{a}{d}$ and $\frac{b}{d}$ are relatively prime. By Theorem 1.5.8 there are r and s such that

$$r\frac{a}{d} + s\frac{b}{d} = 1.$$

By multiplying both side by d we get

$$d = ra + sb.$$

□

Corollary 1.5.10. *Let a, b and k be integers. Suppose that a and k are relatively prime and $k|ab$ then $k|b$.*

Proof. By Theorem 1.5.8 there are integers r and s such that

$$ra + sk = 1.$$

By multiplying both sides by b we get

$$rab + rkb = b.$$

Since $k|ab$ we have $k|rab + rkb = b$. □

The following immediately follows from the corollary above.

Corollary 1.5.11. *Let p be prime and a and b two integers. If $p|ab$ then either $p|a$ or $p|b$.*

Definition 1.5.12. The least common multiple of $a_1, a_2....a_n$, denoted by $lcm(a_1, a_2..., a_n)$, is the positive integer c satisfying the following:

(i) $a_i|c$ for every i.

(ii) If c' satisfies condition (i) then $c|c'$.

As if $c|c'$ implies that $c \leq c'$ therefore c is the lowest among the common factors of a_i's, and automatically unique.

Proposition 1.5.13. *For $a, b \in \mathbb{Z}$, we have $gcd(a,b)lcm(a,b) = ab$.*

Proof. Let $gcd(a,b) = d$ then we have $a = d.a'$ and $b = d.b'$ for some a' and b'. We claim that $gcd(a',b') = 1$. If $k|a'$ and $k|b'$ then $dk|a$ and $dk|b$. Since $d = gcd(a,b)$ we have to have $dk|d$ and this implies that $d = 1$.
We claim that $lcm(a,b) = da'b'$. It is obvious that $a|da'b'$ and $b|da'b'$. Now let k be another integer that $a|k$ and $b|k$. By dividing by d we have $a'|\frac{k}{d}$ and $b'|\frac{k}{d}$. Since a' and b' are relatively prime we have that $a'b'|\frac{k}{d}$. By multiplying by d we have $da'b'|k$. So we conclude that $lcm(a,b) = da'b'$. □

Exercise 1.5.14. If $a = p_1^{e_1}p_2^{e_2}\cdots p_k^{e_k}$ and $a = p_1^{f_1}p_2^{f_2}\cdots p_k^{f_k}$ where $e_i, f_i \geq 0$, $i = 1, ..., k$, then

$$gcd(a,b) = p_1^{\min(e_1,f_1)}p_2^{\min(e_2,f_2)}\cdots p_k^{\min(e_k,f_k)}$$

and

$$lcm(a,b) = p_1^{\max(e_1,f_1)}p_2^{\max(e_2,f_2)}\cdots p_k^{\max(e_k,f_k)}$$

A useful concept in dealing with integers is that of congruence , written $\equiv$. Given three integers a, b and c we shall say $a \equiv b \bmod(c)$ if $a - b$ is a multiple of c i.e.if $a - b$ is divisible by c (all in $\mathbb{Z}$) .

Proposition 1.5.15. *Let a, b, c, d, e and k be integers.*

(i) $a \equiv b \bmod(c)$ *and* $b \equiv d \bmod(c)$ *then* $a \equiv d \bmod(c)$.
(ii) If $a \equiv b \bmod(c)$ *then* $ka \equiv kb \bmod(c)$.
(iii) If $ka \equiv kb \bmod(c)$ *and k and c are relatively prime then* $a \equiv b \bmod(c)$.
(iv) If $a \equiv b \bmod(c)$ *then* $a^n \equiv b^n \bmod(c)$ *for all* $n \in \mathbb{Z}^+$.
(v) If $a \equiv b \bmod(c)$ *and* $d \equiv e \bmod(c)$ *then* $ad \equiv be \bmod(c)$ *and* $a \pm d \equiv b \pm e \bmod(c)$.

Proof. $c|a-b$ and $c|b-d$ then by adding we get $c|a-d$ or $a \equiv d \bmod(c)$. This proves part (i). For part (ii), $ka - kb = k(a-b)$ so if $a-b$ is divisible by c, obviously $ka - kb = k(a-b)$ will be divisible by c. For part (iii), we have $c|ka-kb = k(a-b)$ since k and c are relatively prime then by corollary 1.5.10 we conclude that $c|a-b$.
(iv) follows from the fact that $a^n - b^n = (a-b)(a^{n-1} + a^{n-2}b ... + b^{n-1})$. To prove part (v), notice that since $a \equiv b \bmod(c)$, by part (ii) we have $ad \equiv bd \bmod(c)$. Similarly from $d \equiv e \bmod(c)$ (by multiplying both sides by b) follows that $bd \equiv be \bmod(c)$. Now by part (i) we have $ad \equiv be \bmod(c)$.

□

We shall now determine all Pythagorean triples as mentioned in Exercise1.2.1

Theorem 1.5.16. *If (a, b, c) is a Pythagorean triple, namely*

$$a^2 + b^2 = c^2,$$

then there are integers x and y such that

$$a = x^2 - y^2, \quad b = 2xy, \quad c = x^2 + y^2$$

or

$$b = x^2 - y^2, \quad a = 2xy, \quad c = x^2 + y^2$$

(since the role of a and b can be interchanged).

Proof. We can assume that a, b and c are pairwise relatively prime, because if any two of them has a common divisor then other one has to be divisible by that divisor and we can divide all three of them by that divisor to get another Pythagorean triple. By repeating this we a get triple which are pairwise relatively prime. We recall that multiples of a Pythagorean triple are also Pythagorean triples.

If both a and b are even then c has to be even and this contradicts the assumption that they are relatively prime. If a and b both are odd then $a^2 + b^2 \equiv 2 \bmod(4)$ and c has to be even. But if c is even then $c^2 \equiv 0 \bmod(4)$. So it is impossible to have $a^2 + b^2 = c^2$. One of them is odd and the other one is even. Let a be odd and b be even. Therefore c is odd and we have

$$b^2 = c^2 - a^2 = (c-a)(c+a).$$

We claim that $(c-a, c+a) = 2$. It is obvious that $(c-a, c+a)$ is divisible by 2 as c and a both are odd, so their sum and difference are even numbers. So $(c-a, c+a) = 2d$. In order to prove the claim we must show that $d = 1$.

We have $2d|c-a$ and $2d|c+a$. By adding and subtracting we get $2d|2c$ and $2d|2a$, therefore $d|c$ and $d|a$. As we have assumed that $(c, a) = 1$ so we conclude that $d = 1$. Hence we can write $c + a = 2k$ and $c - a = 2l$ where k and l are relatively prime. By substituting these values we get $b^2 = 4kl$. So kl has to be a perfect square and since k and l are relatively prime each of them has to be a perfect square. Let $k = x^2$ and $l = y^2$. So our previous equation reads $c + a = 2x^2$ and $c - a = 2y^2$. By solving for c and a we have $c = x^2 + y^2$ and $a = x^2 - y^2$, and so $b = 2xy$. □

Exercise 1.5.17. We invite the reader to prove that if the prime factorizations of $a_1, a_2..., a_n$ are given then the prime factorization of the $gcd(a_1, a_2..., a_n)$ contains all the primes that are involved in all a_i's with the minimal exponent. Namely if

$$a_i = p_1^{e_{1i}} p_2^{e_{2i}} \cdots p_k^{e_{ki}}, \quad 1 \leq i \leq n$$

then

$$gcd(a_1, a_2..., a_n) = p_1^{e_1} p_2^{e_2} \cdots p_k^{e_k}$$

where

$$e_j = \min(e_{j1}, e_{j2}, \cdots e_{jn}).$$

Similarly the prime factorization of the lcm is given by taking the maximal exponent.

We can now prove the following generalization of the fact that $\sqrt{2}$ is irrational.

Corollary 1.5.18. *Unless $n \in \mathbb{Z}^+$ is a perfect square, $\sqrt{n}$ is irrational. Conversely, of course, if $n \in \mathbb{Z}^+$ is a perfect square, then $\sqrt{n}$ is rational.*

Proof. Suppose $n \in \mathbb{Z}^+$ is not a perfect square. If $\sqrt{n} = \frac{a}{b}$, where we may, by cancelling out any common factors, assume that a and b have no common factors, then squaring and multiplying by b^2 gives $b^2 n = a^2$. If a prime p divides n, then it divides $b^2 n$ i.e., a^2. Hence, by a corollary above, p must divide a. On the other hand, if a prime p divides a, since it can't divide b and therefore also b^2, p must divide n. Thus the prime divisors on n and a are the same. Let $\{p_1, \ldots, p_k\}$ denote these prime divisors. Then $n = p_1^{e_1} \cdots p_k^{e_k}$ and $a = p_1^{f_1} \cdots p_k^{f_k}$. Since $b^2 n = a^2$ and none of the p_i divides b we see by uniqueness of prime factorization that $e_i = 2f_i$ for each $i = 1 \ldots k$. Since each e_i is even, n is a perfect square. This contradiction proves the corollary. □

In fact this result and its proof can easily be generalized as follows below and is left to the reader as an exercise. We shall formulate an even more general version of this result in section 1.7.

Corollary 1.5.19. *Unless $k \in \mathbb{Z}^+$ is a perfect* n*th power, the* n*th root of k is irrational. Conversely, of course, if $n \in \mathbb{Z}^+$ is a perfect* n*th power, then the* n*th root of k is rational.*

1.6 The Complex Numbers

Notice that it follows from our axioms for an ordered field that for any $x \in \mathbb{R}$, $x^2 \geq 0$. To see this, observe that it is clearly true if $x \geq 0$, while if $x < 0$ all is well since $(-x)^2 = x^2$. From this we see that in spite of the fact that the coefficients are real (indeed integers) there is no *real* solution to the equation $x^2 + 1 = 0$. Just as we have enlarged our domain in order to solve other equations that we were interested in, we now do the same for this one. Fortunately this process will not be endless, for later we shall see that the buck stops here!

We shall call a solution to this later equation i. Thus $i^2 = -1$. Of course, we will also require that $(-i)^2 = -1$. Numbers such as $\pm i$, in fact those of the form bi where $b \in \mathbb{R}$, are called *pure imaginary* numbers. A complex number is, by definition, anything of the form $z = a + bi$, where a and b are real. We refer to a and b, respectively, as the real and imaginary part of z. Thus two complex numbers are the same if and only if they have the same real and imaginary parts. We sometimes also refer to a and b as the *coordinates* of z. The set of all complex numbers is denoted by $\mathbb{C}$.

We add and multiply in $\mathbb{C}$ as follows: If $z = a + bi$ and $w = c + di$, then $z+w = a+c+(b+d)i$ and $zw = ac-bd+(ad+bc)i$. Notice that if we want the distributive law to hold and $i^2 = -1$, then this definition for multiplication is forced on us. As an exercise show that the definition of addition is according to the *parallelogram law* for adding of *vectors* in the plane. Later we shall see the geometric significance of multiplication in $\mathbb{C}$. Except for showing that each $z \neq 0$ has a multiplicative inverse, which will be done in a moment, it is not difficult to check that $\mathbb{C}$ is a field. The remainder of that verification will also be left as an exercise. Observe that if z and w are real (b and d both equal 0), then this addition and multiplication replicate that of the real numbers, so $\mathbb{R}$ is what is called a *subfield* of $\mathbb{C}$. In fact, the reader should check that the same is true in all the extensions we have constructed (from $\mathbb{Z}^+$, to $\mathbb{Z}$, to $\mathbb{Q}$, to $\mathbb{R}$, and finally to $\mathbb{C}$).

For a complex number $z = a + bi$, we define its conjugate $\bar{z} = a - bi$. Then it is a direct verification that $\bar{\bar{z}} = z$ for all $z \in \mathbb{C}$. $\bar{z} = -z$ if and only if z is pure imaginary, $\bar{z} = z$ if and only if z is real and for all complex numbers z and w we have $\overline{(zw)} = \bar{z}\bar{w}$. If $|z|$ is defined by $\sqrt{x^2 + y^2}$, then $z\bar{z} = |z|^2$. Also $|z| = 0$ if and only if $x^2 + y^2 = 0$. But since the square of a real number is always greater than or equal to zero, a sum of squares of real numbers can only be zero if each of the numbers is itself zero; thus in this case, if and only if both x and $y = 0$, that is, if and only if $z = 0$. It follows that if $z \neq 0$, then $z\frac{\bar{z}}{|z|^2} = 1$. Thus z is invertible and $z^{-1} = \frac{\bar{z}}{|z|^2}$.

Let $z \neq 0 \in \mathbb{C}$ and notice that $\frac{z}{|z|}$ always has norm 1. Write $z = |z|\frac{z}{|z|}$. This is called the *polar decomposition* of z. The angle θ, in radians, between the line from the origin through $\frac{z}{|z|}$ (a point on the unit circle) and the positive x axis is called *the argument of* z, written arg(z), while $|z| > 0$ is called the *modulus*. Clearly, $|z|$ is uniquely determined by z and arg(z) is uniquely determined up to an integer multiple of 2π. In fact, because of the definitions of sin and cos we see that

$$\frac{z}{|z|} = \cos\theta + i\sin\theta.$$

This equation shows that conversely, $|z|$ and arg(z) uniquely determine

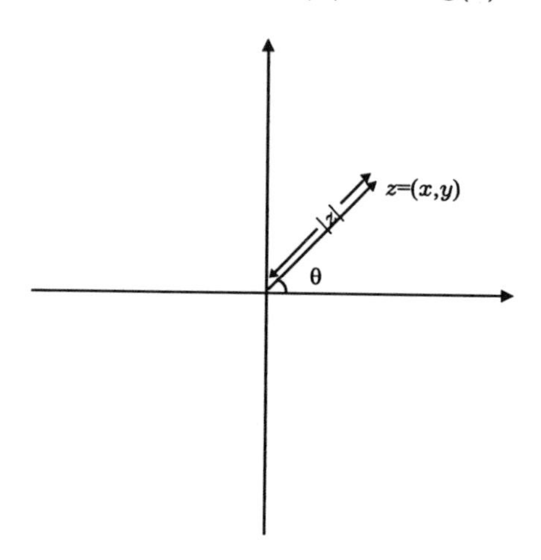

Figure 1.10

z. Here we write $z = |z|(\cos\theta + i\sin\theta)$. [5] We now prove (*DeMoivre's theorem*)

$$|zw| = |z||w|$$

and

$$\arg(zw) = \arg(z) + \arg(w).^6$$

(That is, in multiplying complex numbers, the moduli multiply and the arguments add). In doing so we may clearly assume that both z and w are $\neq 0$, for the first of these equations is trivially satisfied, if one of these is zero, while the second is meaningless.

[5] Later we shall see a more efficient way of organizing this.

[6] Actually, $\arg(zw) = \arg(z) + \arg(w) (\bmod 2\pi)$.

Given two complex numbers of unit length, $\cos\theta + i\sin\theta$ and $\cos\phi + i\sin\phi$, making use of the usual trigonometric identities, we multiply them and get $(\cos\theta\cos\phi - \sin\theta\sin\phi) + i(\cos\theta\sin\phi + \sin\theta\cos\phi) = \cos(\theta+\phi) + i\sin(\theta+\phi)$. In particular, the product of two complex numbers of unit length also has unit length and $\arg(zw) = \arg(z) + \arg(w)$. Now suppose we also look at the polar decomposition of $w = |w|(\cos\phi + i\sin\phi)$. Then

$$zw = |z|(\cos\theta + i\sin\theta)|w|(\cos\phi + i\sin\phi) = |z||w|(\cos(\theta+\phi) + i\sin(\theta+\phi)).$$

Since $|z||w| > 0$ and $\arg(zw) = \arg(z) + \arg(w)$, we see by uniqueness that $|zw| = |z||w|$.

An important feature of $\mathbb{C}$ (called the *Fundamental Theorem of Algebra*) is that any polynomial equation $p(z) = 0$ of any positive degree with complex coefficients always has all its roots in $\mathbb{C}$. For the moment we shall content ourselves by proving the Fundamental Theorem of Algebra in an interesting special case; namely, by showing that over $\mathbb{C}$ we can always solve the equation $z^n - a = 0$, where $a \in \mathbb{C}$. Later on, in the chapter on topology of the plane we will prove it in general. The proof will use other methods, but will ultimately rely on facts about the polynomial z^n. Now assume $a \neq 0$ (we can always find the n*th* roots of 0). Its polar decomposition is given by $a = |a|(\cos\theta + i\sin\theta)$. By DeMoivre's theorem

$$z = |a|^{\frac{1}{n}}(\cos(\frac{\theta}{n}) + i\sin(\frac{\theta}{n}))$$

is an n*th* root of a. Let

$$\omega = \cos\frac{2\pi}{n} + i\sin\frac{2\pi}{n}.$$

By DeMoivre's theorem, we get $\omega^n = 1$, but for $j = 1, \ldots, n-1$ we see that $\omega^j \neq 1$ since $0 < \frac{2\pi j}{n} < 2\pi$ and so it doesn't even have the right argument to be 1. It follows that the set of all powers $S = \{\omega, \omega^2, \ldots \omega^n\}$, called the n*th roots of unity* are distinct. For if $\omega^j = \omega^k$ where $1 \leq j < k \leq n$, then we would get $\omega^{k-j} = 1$, a contradiction, since $1 < k - j < n$. Moreover, for $j = 1, \ldots, n$ each ω^j has the property that

$(\omega^j)^n = (\omega^n)^j = 1^j = 1$. Since the n*th* roots of unity are distinct, it follows that $\{z\omega, z\omega^2, \ldots z\omega^n\}$ are also distinct and therefore consist of n complex numbers. Also $(z\omega^j)^n = z^n(\omega^j)^n = a1 = a$. Thus each one of these is an n*th* root of a. Since, as we shall see shortly, a polynomial equation of degree n has at most n roots, this must be all of them.

Exercise 1.6.1.

1. Draw a picture of the n*th* roots of unity. These provide the solution to the very special case $z^n - 1 = 0$ of the Fundamental Theorem of Algebra.

2. Use this picture to show that the sum of the n*th* roots of unity is always zero.

Definition 1.6.2. An n*th* root of unity z is called *primitive*, if the power z^m is never 1 for $m = 1, \ldots, n-1$.

Use the fundamental theorem of arithmetic above and induction on n to do the following exercises:

Exercise 1.6.3.

1. Show if n is prime, then every n*th* root of 1 is primitive.

2. Let $n > 1$, P be the primitive n*th* roots of 1 and $p(z) = \Pi_{\sigma \in P}(z - \sigma)$. Show p is a polynomial with *integer* coefficients.

3. Let $n > 1$, and $p(z)$ be as above. Then $p(z)$ divides $\frac{z^n - 1}{z^d - 1}$, where d is a divisor of n in $\mathbb{Z}$.

The proof of the Fundamental Theorem of Algebra in its most general form cannot be given at this level, and so will have to be deferred until Chapter 3. However, assuming the truth of this result we can investigate the nature of the roots of a polynomial equation $p(x) = a_n x^n + a_{n-1}x^{n-1} + \ldots + a_0 = 0$ with real coefficients. Let z be a complex, non real root of the equation. Taking conjugates, and using the properties of conjugation mentioned above, we get $a_n(\bar{z})^n + a_{n-1}(\bar{z})^{n-1} + \ldots + a_0 = 0$, that is $p(\bar{z}) = 0$. Thus the non real roots must occur in conjugate pairs.

We now consider the most general quadratic equation

$$ax^2 + bx + c = 0,$$

which we imagine for the moment has real coefficients. Later we will consider the possibility of complex coefficients. Of course the leading coefficient $a \neq 0$ since otherwise we would have a linear equation and not a quadratic equation. Writing it as $ax^2 + bx = -c$ and dividing by a gives $x^2 + \frac{b}{a}x = -\frac{c}{a}$. Now we add $\frac{b^2}{4a^2}$ to both sides. The effect of this is to make the left hand side a *perfect square*. This process is known as *completing the square*. It tells us $x^2 + \frac{b}{a}x + \frac{b^2}{4a^2} = (x + \frac{b}{2a})^2$, while $\frac{b^2}{4a^2} - \frac{c}{a} = \frac{b^2-4ac}{4a^2}$. Thus we get

$$(x + \frac{b}{2a})^2 = \frac{b^2 - 4ac}{4a^2}.$$

We define the *discriminant* $\Delta = b^2 - 4ac$. Now $4a^2 = (2a)^2$ is positive. If $\Delta > 0$, then we can take square roots in $\mathbb{R}$. Solving for x we get

$$x = \frac{-b \pm \sqrt{b^2 - 4ac}}{2a}$$

Similarly, if $\Delta = 0$, we get the same thing except now both roots are equal (to $\frac{-b}{2a}$). If on the other hand $\Delta < 0$, then we can take the square root $\pm i\sqrt{-(\Delta)}$ in $\mathbb{C}$ and get two distinct complex conjugate solutions to the quadratic equation,

$$\frac{-b \pm \sqrt{-\Delta}i}{2a}.$$

This shows that the only possible solutions are the ones mentioned above. To see that they are indeed solutions we must substitute them into the original equation and see if they satisfy it. We leave this latter task as an exercise. Notice we have proven that there are at most two solutions to any quadratic equation, and that any non real solutions at least lie in $\mathbb{C}$ and are conjugate. We leave it, also as an exercise, for the reader to show that the sum of the roots is $-\frac{b}{a}$ and the product of the roots is $\frac{c}{a}$.

Exercise 1.6.4. The *golden mean* of the Greeks is the ratio of (smaller) side y, to base x, of a rectangle with the property that

$$\frac{x}{y} = \frac{y}{x+y},$$

from which it follows that $x(x+y) = y^2$. Solving this quadratic equation for x, and taking into account that all quantities here are positive, verify that the golden mean is

$$\frac{x}{y} = \frac{\sqrt{5}-1}{2}.$$

Now, what problems would arise if the coefficients of our general quadratic equation were in $\mathbb{C}$ rather than $\mathbb{R}$? The answer is none. The argument would be just the same only simpler, since there would be no case distinctions. Whatever the value of $\Delta \in \mathbb{C}$, we have seen that we can find its (two) square roots in $\mathbb{C}$. Hence the solutions to the quadratic equation are:

$$z = \frac{-b \pm \sqrt{\Delta}}{2a}.$$

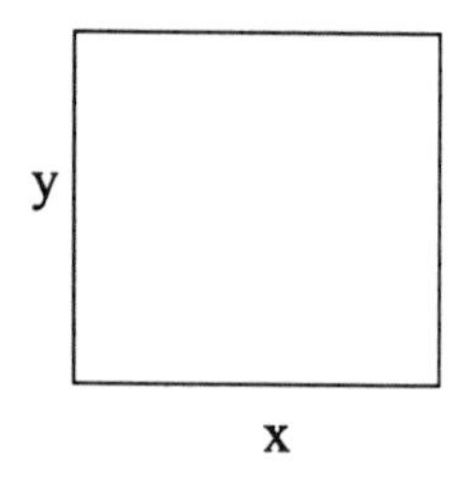

Figure 1.11

All other features remain the same as in the real case.

We conclude this section with the computation of the sum of the terms of a *geometric progression* which is defined as follows. Let a_0 and $z \neq 1 \in \mathbb{C}$ and form the following:

$$a_0, a_0 z, a_0 z^2, \ldots \tag{1.6}$$

This is completely analogous to an arithmetic progression, but with addition replaced by multiplication. In fact our derivation of the formula for the sum of the terms of an arithmetic progression would work just as well if the terms were complex numbers rather than merely integers. (Try this as an exercise). We just didn't want to wait to introduce this concept. As in the case of an arithmetic progression let $\Sigma(k) = a_0 + a_0 z + \ldots + a_0 z^k$. Then $z\Sigma(k) = a_0(z + z^2 + \ldots z^{k+1})$. Therefore when we subtract there is massive cancellation; $\Sigma(k) - z\Sigma(k) = a_0(1 - z^{k+1})$. Solving, we get $\Sigma(k) = \frac{a_0(1-z^{k+1})}{1-z}$.

We have already shown if $|z| > 1$ then $\lim_{n \to +\infty} |z|^n = +\infty$. Hence if $|z| < 1$, it follows that $\lim_{n \to +\infty} |z|^n = 0$. From this we shall see that

Theorem 1.6.5. *If* $|z| < 1$, *then the* infinite series, $a_0 + a_0 z + \ldots + a_0 z^k + \ldots$, *usually written* $\Sigma_{k=0}^{\infty} a_0 z^k$, *and which is called the* geometric series, converges *to* $\frac{a_0}{1-z}$.

Proof. Let $S_n = \sum_{k=0}^{n} a_0 z^k$. We have

$$zS_n = \sum_{k=1}^{n+1} a_0 z^k.$$

then as above

$$S_n = \frac{a_0 - a_0 z^{n+1}}{1 - z}.$$

We have $S_n = \frac{a_0}{1-z} - (\frac{a_0}{1-z}) z^n$. The last term converges to zero as $n \to \infty$. Hence $\sum_{k=0}^{\infty} a_0 z^k = \frac{a_0}{1-z}$.

□

We can apply this last fact to distinguish rational numbers from irrational ones within $\mathbb{R}$. Let the decimal expansion of a real number be $x = n.a_1 a_2 \ldots a_n \ldots$ and let us say that this expansion is *periodic* if, after a finite number of places, say k, it repeats. That is, there is a $\nu \in \mathbb{Z}^+$ so that starting from the kth place it reads

$$a_k a_{k+1} \ldots a_{k+\nu-1} a_k a_{k+1} \ldots a_{k+\nu-1} \ldots,$$

We will now prove the following

Theorem 1.6.6. *Each $x \in \mathbb{Q}$ has a periodic decimal expansion. Conversely, if a real number x has a periodic decimal expansion, then $x \in \mathbb{Q}$.*

Proof. Indeed, suppose the real number x has a periodic decimal expansion. Then after subtracting off a rational number $r = n.a_1 a_2 \ldots a_{k-1}$ we see that

$$10^{k-1}(x - r) = .a_k a_{k+1} \ldots a_{k+\nu-1} a_k a_{k+1} \ldots a_{k+\nu-1} \ldots.$$

Hence, there is an $s \in \mathbb{Z}^+$ such that $10^{k-1}(x-r) = s(10^{-\mu} + 10^{-2\mu} + \cdots)$ Since $\frac{1}{10^\mu} < 1$, $\frac{10^{k-1}(x-r)}{s}$ is a geometric series whose sum is $\frac{10^{-\mu}}{1-10^{-\mu}}$, which is clearly rational. Hence, so is x.

Conversely, given a rational number $\frac{m}{n}$ we can compute its decimal expansion by long division, i.e., using the Euclidean algorithm above. If at any stage the remainder $r = 0$, then the resulting decimal expansion will have only zeros from that point on and thus is periodic. The only other possibility is that $1 \leq r \leq n - 1$. Thus there are $n - 1$ possible values of r. In any interval of n places one of these will have to repeat. Then the process of long division will just give the same result again. Thus the decimal expansion is periodic of period at most n. □

The idea that when there are fewer possible values than places then one of these values will have to repeat is a typical use of the so called *pigeon hole principle* of a branch of mathematics called *combinatorics*.

We can now construct many new irrational numbers. For example, the number .1010010001... is clearly irrational since it can have no period. (Such numbers were first identified by Liouville). We also know that if $n \in \mathbb{Z}^+$ is not a perfect square, then the decimal expansion of the $\sqrt{n}$ never repeats.

1.7 Polynomials and Other Analogues of the Integers

We now apply this same concept of long division to polynomials $p(x)$ with real or complex coefficients, or indeed coefficients from any field to prove the analogue of the Euclidean algorithm for polynomials and as usual the r defined below is called the remainder.

Theorem 1.7.1. *Given polynomials $b(x)$ and $a(x)$, we can find polynomials $p(x)$ and $r(x)$ satisfying $b(x) = p(x)a(x) + r(x)$, where* $\deg r < \deg a$.

Proof. Let $b(x) = b_n x^n + b_{n-1}x^{n-1} + \ldots + b_0$ and $a(x) = a_m x^m + a_{m-1}x^{m-1} + \ldots + a_0$, where we may assume that $m \leq n$; otherwise there is nothing to prove. Then b_n and $a_m \neq 0$. Clearly, $b(x) - \frac{b_n}{a_m} x^{n-m} a(x)$ has degree smaller than $\deg b$. Therefore, we can continue and complete the argument by induction on the degree of b. □

This statement is the exact analogue of the Euclidean algorithm in $\mathbb{Z}$. Hence, reasoning exactly as in $\mathbb{Z}$, we also get the analogues (see below) of all its consequences. As a test of your understanding of these facts in $\mathbb{Z}$ it would be worthwhile proving their analogues for polynomials. Here a polynomial is called a *prime polynomial* if it can't be factored into polynomials of positive degree.

Corollary 1.7.2. *If a prime polynomial $p(x)$ divides $a(x)b(x)$, where $a(x)$ and $b(x)$ are polynomials, then $p(x)$ divides $a(x)$ or $b(x)$ (or both). Moreover, any polynomial can be factored uniquely into prime polynomials.*

Now suppose $b(x)$ is a polynomial and $x - c$ is a factor, that is $b(x) = p(x)(x - c)$. Then clearly c is a root of $b(x) = 0$. Conversely, if c is a root of $b(x) = 0$, applying the Euclidean algorithm we get $b(x) = p(x)(x-c)+r$. Since r has degree 0, it is a constant. Evaluating at $x = c$ tells us the constant is 0 and therefore $b(x) = p(x)(x-c)$. Thus c is a root of a polynomial if and only if $x - c$ is a factor. In particular a polynomial equation of degree n can have at most n roots, Now it follows from the Fundamental Theorem of Algebra that any polynomial $p(z)$ of degree n and leading coefficient c with coefficients from $\mathbb{C}$ factors as $p(z) = c(z - \alpha_1)\ldots(z - \alpha_n)$. If the coefficients happened to be in $\mathbb{R}$, the non real roots occur in conjugate pairs. If $a + bi$ and $a - bi$ are two such roots, then the product of the corresponding linear factors is $(z - a + bi)(z - (a - bi)) = z^2 - 2az + a^2 + b^2$. The discriminant of this quadratic polynomial (with real coefficients) is $-b^2$, which is clearly negative, since $b \neq 0$ is real. Thus we have proven that a polynomial $p(x)$ with real coefficients factors into a product of linear factors of the form $x - a$ where $a \in \mathbb{R}$, together with some quadratic polynomials each of which have negative discriminant. As an exercise find which polynomials with complex or real coefficients, respectively, are prime.

We can also do a similar thing for the *Gaussian integers*. These are the complex numbers of the form $\mathcal{G} = \{n+mi : n, m \in \mathbb{Z}\}$. We shall see that this is a system of "numbers" similar in character to that of the integers and polynomials. We define the *norm* of a Gaussian integer, or indeed any complex number, by the formula $N(n + mi) = n^2 + m^2$. In other words, $N(n+mi) = |n+mi|^2$. In particular, $N(z)N(w) = N(zw)$ and $z = 0$ if and only if $N(z) = 0$. The norm will play a role similar to that of the degree of a polynomial. Now given $a, b \in \mathcal{G}$, with $a \neq 0$ there are q, $r \in \mathcal{G}$ such that $b = aq + r$ and $N(r) < N(a)$. This is the exact analogue of the Euclidean algorithm.

To see that the Euclidean algorithm is true for $\mathcal{G}$, divide b by a in $\mathbb{C}$. The quotient $\frac{b}{a}$ will most likely not have integer coordinates, but clearly there will be a point q with integer coordinates such that $N(\frac{b}{a} - q) \leq (\frac{1}{2})^2 + (\frac{1}{2})^2 = \frac{1}{2}$. Hence $N(b - aq) = N(a)N(\frac{b}{a} - q) < \frac{1}{2}N(a) < N(a)$. Letting $r = b - aq$, we see that r is a Gaussian integer whose norm is less that of a. We now define primes of $\mathcal{G}$ to be those Gaussian integers which can't be factored except when one of the factors has norm 1. But this is tricky; for example, 3 is prime in $\mathcal{G}$ (prove), but 5 is not, since $5 = (2+i)(2-i)$. We shall soon see which primes in $\mathbb{Z}^+$ are actually primes in $\mathcal{G}$.

Exercise 1.7.3. Find which elements in $\mathcal{G}$ have norm 1 and which elements in $\mathcal{G}$ actually have inverses in $\mathcal{G}$.

Exercise 1.7.4. Use the same method in the ring $\mathbb{C}[x]$ or $\mathbb{R}[x]$.

Consequently, we can conclude (exercise): If a prime $p \in \mathcal{G}$ divides ab, then p divides a or b (or both). Moreover, any $a \in \mathcal{G}$ can be factored uniquely into primes of $\mathcal{G}$.

Finally, we apply some of these facts about the Gaussian integers to prove a result concerning the integers themselves. This will tell us which prime integers are a sum of squares of two integers and eventually will lead us to a famous theorem of Fermat. For example, 3 clearly cannot be written as a sum of two squares, while $2 = 1^2 + 1^2$ and $5 = 2^2 + 1^2$. More generally, if p is a prime of $\mathbb{Z}$, but not of $\mathcal{G}$, then applying conjugation to the prime factorization yields another prime factorization of p. Hence by uniqueness of prime factorization the factors not in $\mathbb{Z}$ must occur in conjugate pairs, say $n + mi$ and $n - mi$. But then $(n+mi)(n-mi) = n^2 + m^2$ must divide p. Since $n^2 + m^2$ is an integer and p is a prime of $\mathbb{Z}$, $p = n^2 + m^2$. Thus any prime of $\mathbb{Z}$ which is not a prime of $\mathcal{G}$ is a sum of squares of two integers. Conversely, if the prime $p \in \mathbb{Z}$ is of the form $p = n^2 + m^2$, then $p = (n+mi)(n-mi)$ and so p can't be a prime in $\mathcal{G}$, unless $N(n \pm mi) = n^2 + m^2 = 1$, that is, one of these integers is 0 and the other is ± 1. In this case, $p = 1$ and so is not a prime. Thus we have proven the following:

Corollary 1.7.5. *A prime of $\mathbb{Z}$ is a sum of squares of two integers if and only if it is not a prime of $\mathcal{G}$.*

Now it is reasonable to ask if this situation can be described completely in terms of integers. The answer is yes and this can be done in a very useful way provided by the following result, first stated by Fermat, but actually proven by Euler. Although we are not in a position to prove it in this section, we will in the succeeding one. This result has applications to determining whether a large integer is prime or not.

Theorem 1.7.6. *A prime $p > 2 \in \mathbb{Z}$, is a sum of squares of two integers if and only if it has the form $p = 4n + 1$, where $n \in \mathbb{Z}$.*

Let $\mathbb{A}$ denote the *algebraic numbers.* These are the complex numbers which are solutions to polynomial equations $p(x) = 0$, where the polynomial has integer coefficients and leading coefficient 1. For example the n*th* root of the integer k is an algebraic number since $x^n - k$ is a polynomial of the type just mentioned.

We now come to a considerable generalization of the "irrationality" results proved earlier.

Theorem 1.7.7. $\mathbb{A} \cap \mathbb{Q} = \mathbb{Z}$.

Thus we see that unless k is a perfect n*th* power, the n*th* root of k must be irrational.

Proof. Of course $\mathbb{Z} \subseteq \mathbb{A} \cap \mathbb{Q}$. To prove the converse let x be a solution to $p(x) = x^n + a_1 x^{n-1} + \ldots a_n = 0$ and suppose $x = \frac{a}{b}$, where a, $b \neq 0 \in \mathbb{Z}$. If a and b have any common factors cancel them. Thus we may assume they have no common factors. Substituting and clearing denominators yields $a^n + a_1 b a^{n-1} + \ldots a_n b^n = 0$, an equation in $\mathbb{Z}$. Since b divides all terms save the last it must also divide a^n. Let p be a prime dividing b. Then p must divide a^n. Hence, as we saw above, p must actually divide a. But then both a and b would have p as a common factor. Since this is impossible it follows that b has no prime divisors at all. Therefore $b = \pm 1$ and $\frac{a}{b} \in \mathbb{Z}$. □

As we saw $\mathcal{G}$ the Gaussian integers satisfy something quite analogous to the division algorithm of the integers, or the polynomials in one variable with coefficients from a field. Consequently it shares many properties of these commutative rings with identity, including unique prime factorization. As an example the reader is invited to verify that the same is true of the ring

$$R_2 = \{n + m\sqrt{-2} : n, m \in \mathbb{Z}\}$$

with the usual properties of addition and multiplication. The method for doing this is quite analogous with that of $\mathcal{G}$. Notice that here if

$x = n + m\sqrt{-2}$ and we define $N(x) = n^2 + 2m^2 \in \mathbb{Z}$, then $N(xy) = N(x)N(y)$ for all x and $y \in R_2$ and $n^2+2m^2 = (n+m\sqrt{-2})(n-m\sqrt{-2})$. We invite the reader to find all invertible elements in R_2. Then proceed to verify the division algorithm as in the case of $\mathcal{G}$ using the diagram below.

We now give an example of what would seem to be a similar algebraic structure to that of the Gaussian integers, but which actually has very different properties. Namely,

$$R_3 = \{n + m\sqrt{-3} : n, m \in \mathbb{Z}\}$$

these numbers don't have unique factorization into primes. In particular, the analogue of the division algorithm cannot be satisfied here. One verifies, just as with R_2, that $N(xy) = N(x)N(y)$ for all x and $y \in R_3$ and $n^2 + 3m^2 = (n + m\sqrt{-3})(n - m\sqrt{-3})$. From this it follows that the only invertible elements are ± 1. However $(1+\sqrt{-3})(1-\sqrt{-3}) = 1+3 = 4 = 2 \cdot 2$. Since none of the terms $1 + \sqrt{-3}$, $1 - \sqrt{-3}$, 2 is invertible and each is prime in R_3, this violates uniqueness of prime factorization. As an exercise we invite the reader to explain what goes wrong with the argument which worked so successfully in $\mathcal{G}$ and in R_2 when we try to use it in R_3?

We now turn to the Chinese Remainder Theorem which seems to have been first posed in 4 CE.

Theorem 1.7.8. *Let $\{a_1, \ldots, a_k\}$ be a set of k integers which are pairwise relatively prime.*

Given integers $\{x_1, \ldots, x_k\}$ one can simultaneously solve the system of congruences $x \equiv x_i \operatorname{mod}(a_i)$, $i = 1, \ldots, k$ for $x \in \mathbb{Z}$.

In particular, it follows from the Chinese Remainder Theorem that if $n \in \mathbb{Z}$ and $n = p_1^{e_1} \cdots p_k^{e_k}$ is its prime factorization, then

$$\mathbb{Z}/(n) \cong \Pi_{i=1}^k \mathbb{Z}/(p_i^{e_i}).$$

For we clearly have $p_i^{e_i}$ and $p_j^{e_j}$ have no common prime divisors if $i \neq j$.

Proof. For each $i = 1, \ldots, k$ let $b_i = \Pi_{j \neq i} a_j$. Then, for each i, a_i and b_i are relatively prime. For suppose for some i there was a prime number p so that $p|a_i$ and $p|b_i$. Then $p|\Pi_{j\neq i}a_j$. Hence by Theorem 1.5.3 $p|a_j$ for some $j \neq i$. This contradicts our assumption that the a_i are pairwise relatively prime. Therefore, by Theorem 1.5.8, for each $i = 1, \ldots, k$ there are integers α_i and β_i so that $\alpha_i a_i + \beta_i b_i = 1$. Let $\gamma_i = 1 - \alpha_i a_i$ where $i = 1, \ldots, k$. Then since $\gamma_i \in (a_j)$ for all $j \neq i$ we have $\gamma_i \equiv 0 \bmod(a_j)$ for $j \neq i$, whereas

$$\gamma_i \equiv 1 - \alpha_i a_i \equiv 1 - \alpha_i a_i \equiv 1 - 0 \equiv 1 \bmod(a_j).$$

Thus $\gamma_i \equiv 1 \bmod(a_j)$ for $i \neq j$ and $\gamma_i \equiv 0 \bmod(a_j)$ otherwise. Now given integers $x_1, x_2, ..., x_k$, for $x = \sum_{i=1}^{k} x_i\gamma_i$ we have

$$\Sigma_{i=1}^{k} x_i\gamma_i \equiv x_j\gamma_j \bmod(a_j) \equiv x_j.$$

So x is a solution to the problem.If y is another solution we have,

$$x - y \equiv 0 \bmod a_i, \quad i = 1, 2, ...k$$

or $a_i|x - y$, for $i = 1, 2..., k$. Since a_i's are relatively prime we conclude that $a_1a_2...a_k|x - y$, or in other words $x - y \in (a_1a_2...a_k)$. So the difference of x and y is a multiple of $a_1a_2...a_k$. Conversely, if x is a solution to the problem and we add a multiple of $a_1a_2...a_k$ to x and we get a new solution. □

1.8 The Algebra on a Dial

We now consider the algebra of numbers on a dial. To make this quite concrete we shall assume the dial resembles a standard clock with twelve marks placed in the usual way except that it has only one hand, the hour hand. (However, we observe that everything we shall say here would apply as well if the number of markings on the dial were any positive integer greater then or equal to two). On this dial when we add "hours" we move in a clockwise direction and when we subtract "hours" we move in a counter clockwise direction. For example, $8 + 7 = 3$ and $6 - 10 = 8$.

Notice that when any addition (or subtraction) problem is done which results in the answer 12 the result is 0. Actually, the same may be said of 24 or -12, or indeed any integer multiple of 12. Thus all integer multiples of 12 are equivalent to 0 and, in fact, $x \equiv y$ if and only if $x - y$ is divisible by 12. Notice $x + 0 = x$ for any x.

In this way we have a set of 12 elements, usually written $\mathbb{Z}_{12}$ which has the following properties: Any two elements x and y when added or subtracted result in a definite element on the dial. For any two elements $x + y = y + x$, and for any three elements $(x + y) + z = x + (y + z)$. These last two are respectively the commutative and associative laws for addition. Since as we already noted $x + 0 = x$ this means that 0 is the additive identity. Such a system is usually called an Abelian group.

Now we can also multiply these dial numbers. For example, $5 \cdot 7 = 11$ $3 \cdot 4 = 0$ and $1 \cdot x = x$ for any x. The properties of this multiplication are the following: Any two elements x and y when multiplied result in a definite element on the dial. For any two elements $xy = yx$, and for any three elements $(xy)z = x(yz)$. Just as before, these last two are respectively called the commutative and associative laws for multiplication. Since as we already noted, $1x = x$ this means that 1 is the multiplicative identity. Multiplication and addition are connected by the fact that $(x + y)z = xz + yz$. This last is called the distributive law.

To reiterate the situation, described earlier, a set R with an addition $+$ and a multiplication $\cdot$ satisfying the above conditions and which also satisfies the distributive law is called a *commutative ring with identity*. In such a ring, we shall call an element x *invertible* if there is an element of the ring y in the ring such that $xy = 1$. It is easy to see that y is uniquely determined by x. For this reason we can write $y = x^{-1}$.

Thus $\mathbb{Z}_{12}$ is a commutative ring with identity. The same may be said for $\mathbb{Z}_n$, where $n \geq 2$ as well as $\mathbb{Z}$ itself. Now one difference between some of the $\mathbb{Z}_n$ and $\mathbb{Z}$ is that in the former one can have a product of two nonzero elements equals to zero. Such things are called *zero-divisors*. As we saw, this can't happen in $\mathbb{Z}$. In fact as we shall see later, $\mathbb{Z}_n$ has no zero divisors if and only if n is a prime number.

As before, a commutative ring with identity is called a field if in addition to all the properties mentioned just above it is also the case that any non zero element x has a multiplicative inverse, x^{-1}. Thus $xx^{-1} = 1$. We leave as an exercise to the reader to explain why zero can't have a multiplicative inverse. For example, $\mathbb{Z}$ is not a field, but $\mathbb{Q}$, $\mathbb{R}$ and $\mathbb{C}$ are fields. If the integer n isn't prime, (e.g. $12 = 3 \cdot 4 = 0$), then clearly $\mathbb{Z}_n$ has zero divisors and so can't be a field. On the other hand (see theorem 1.5.8) if $n = p$, a prime, then $\mathbb{Z}_n$ is a field. To see this, suppose that p is a prime and

$$x \neq 0 (\operatorname{mod} p).$$

Since p is a prime, then $(x, p) = 1$. So there exit integers a and b such that

$$ap + bx = 1,$$

thus

$$bx \neq 1 (\operatorname{mod} p).$$

or in other words b inverts in $\mathbf{Z}_p$.

Therefore $\mathbb{Z}_n$ is a field if and only if n is a prime number.

Chapter 2

Groups, Finite Fields and Linear Algebra

2.1 Introduction to Groups and Their Actions

Groups and their actions have proven central to the study of both geometry and algebra. This was first fully realized by the late 19th century mathematician F. Klein in his famous *Erlangen* Program. Groups have also had a great unifying effect on mathematics causing the 20th century mathematician A. Weil to make the statement "Il n'est qu'une mathématique" (there is only one mathematics), and thus forever changing the spelling of the word in French from *mathématiques* to *mathématique*. One good effect of the unifying of mathematics is not so much that it is easier to learn then heretofore, but that *for the same effort one can learn much more!*

We begin with the definition of a *group*. This is a set G, together with a *binary operation* $\cdot$, satisfying the following properties (or *axioms*) given below. Usually we write $(G, \cdot)$. Before stating these properties we remark that a binary operation simply means one involving two elements of G. Sometimes we will suppress the $\cdot$. The following axioms are not absolutely minimal, but are a comfortable set to work with.

1. If g and $h \in G$ then $gh \in G$ (closure).

2. For g, h and $k \in G$, $(gh)k = g(hk)$ (associative law).

3. There exists a $1_G \in G$ satisfying $1_G g = g 1_G = g$ for all $g \in G$. Henceforth, we shall write 1 for 1_G if there is not danger of confusion.

4. For each $g \in G$ there is $g^{-1} \in G$ such that $gg^{-1} = g^{-1}g = 1$.

We remark that the 1 and g^{-1} in the last two statements are unique. To see this, suppose $ag = g$ for some $a \in G$ and all $g \in G$. Taking $g = 1$ gives $a1 = 1$. But $a1 = a$. Thus, $a = 1$. Similarly, for $g \in G$, if $gh = 1$, then multiplying by g^{-1} gives $g^{-1}(gh) = g^{-1}1$. Hence $(g^{-1}g)h = 1h = h = g^{-1}$. 1 is called *the identity* of G while g^{-1} is called *the inverse* of g.

An important distinction is whether a group is commutative (sometimes called Abelian) or not, that is whether or not $gh = hg$ for all g and $h \in G$. When a group is commutative the operation is often written additively. Another important distinction is whether G has a finite or infinite number of elements. For example as above, we denote by S_n the set of all permutations of X_n, where by a permutation we mean a 1:1 and onto, or *bijective* mapping of X_n to itself. Let f and g be maps of a set X to itself. Their composition, $f \cdot g$ is defined by $f \cdot g(x) = f(g(x))$, for $x \in X$. If we take for the group operation that of composing permutations, then S_n is a finite group which is not commutative when $n > 2$. S_2 is actually commutative. More generally, if X is any set, its permutations under composition, exactly in the sense used above, is always a group. On the other hand, the additive group $(\mathbb{Z}, +)$ or the multiplicative group $(F \setminus \{0\}, \cdot)$, where F is a field and 0 is the zero element of F, are commutative. This is called the multiplicative group of the field and is often written $F^\times$. As we have seen, examples of infinite fields are $\mathbb{Q}$, $\mathbb{R}$ and $\mathbb{C}$. As we shall see in the next section $\mathbb{Z}_p$, where p is a prime in $\mathbb{Z}$, is a finite field and so is its multiplicative group $(\mathbb{Z}_p^\times, \cdot)$. These statements should be verified by the reader.

Let G be a group and H be a subset of G which contains 1 and is closed under multiplication and the taking of inverses. Then under the (restricted) multiplication, H is itself a group. When this happens we call H a *subgroup* of G. For example, in $(\mathbb{Z}, +)$ if m is any fixed integer and $H = \{km : k \in \mathbb{Z}\}$, then H a subgroup of $(\mathbb{Z}, +)$. If G is any group and $g \in G$, then $\{g^n : n \in \mathbb{Z}\}$ is a subgroup, called the *cyclic subgroup generated by* g. An example of a cyclic subgroup of $\mathbb{C}^\times$ is the group of n*th* roots of unity. An example of a subgroup of $S(X)$, the permutations of a set X, can be gotten by choosing a subset $Y \subseteq X$ and taking for the subgroup H the permutations that stabilize Y, that is that send Y into itself. A smaller subgroup is given by the permutations that actually fix each point $y \in Y$. The reader should check that these are all subgroups or cyclic subgroups, respectively.

Now suppose G is a group and H a subgroup. For each $g \in G$ we call the set $gH = \{gh : h \in H\}$ the *left coset* of H determined by g. Notice that g_1H could equal g_2H even though $g_1 \neq g_2$. In fact, $g_1H = g_2H$ if and only if $g_2^{-1}g_1 \in H$. Notice also that if $g_1H \cap g_2H \neq \phi$, then $g_1H = g_2H$. We denote the set of all left cosets by G/H and call the natural map $\pi : G \to G/H$. We shall look at this construction in more detail in the group $(\mathbb{Z}, +)$.

Notice that G is the disjoint union of cosets, each of which is non empty. Later we shall see this situation in a more general context. Observe that these cosets are all in bijective correspondence since $h \mapsto gh$ is a bijective map from $H \to gH$. As a consequence we get the theorem of Lagrange: The *index of a subgroup* is the number of its cosets.

Theorem 2.1.1. *In a finite group, the order of any subgroup divides the order of the group.*

If G is a group and $g \in G$, we say g has *finite order* if $g^n = 1$ for some $n \in \mathbb{Z}^+$. The smallest such n is called the order of g. Using the Euclidean algorithm if the order of g is m, then the cyclic subgroup generated by g is isomorphic to $(\mathbb{Z}_m, +)$. (This concept will be defined immediately below). Otherwise we say g has *infinite order*. In this case the cyclic subgroup generated by g is isomorphic to $(\mathbb{Z}, +)$. Hence if G is finite, every element has finite order. Since the order of an element

g is the order of the subgroup generated by G we see from the theorem of Lagrange that

Corollary 2.1.2. *The order of any element in a finite group divides the order of the group.*

When should we consider two groups G_1 and G_2 to be essentially the same, meaning exactly the same except perhaps for the names of the players? We shall say that G_1 and G_2 are isomorphic if there is map $f : G_1 \to G_2$ which is bijective satisfying for all $g, h \in G_1$

$$f(gh) = f(g)f(h).$$

We call such an f an isomorphism. If f satisfies this condition, but is not necessarily bijective, then it is called a homomorphism. We shall call $Ker f = \{g \in G_1 : f(g) = 1\}$ the *kernel* of such a homomorphism.

Exercise 2.1.3.

1. Prove if H is a subgroup of G, then $gHg^{-1} = H$ for all $g \in G$ if and only if each left coset of H is a right coset. When this happens we say that H is a *normal subgroup*.

2. Of course, if G is Abelian, then every subgroup is normal. Find an example of a group and a subgroup which is not normal.

3. Show that for a group homomorphism $f(1) = 1$.

4. Prove if H is a normal subgroup of G then G/H can be made into a group and the map $\pi : G \to G/H$ sending $g \mapsto gH$ is a homomorphism whose kernel is H.

5. Suppose $f : G_1 \to G_2$ is a homomorphism. Show that its kernel H is a normal subgroup of G_1, $f(G_1)$ is a subgroup of G_2 and finally that there is an induced isomorphism $G_1/H \to f(G_1)$. This is called the first isomorphism theorem.

6. $\mathbb{R}^\times$ is not isomorphic to $(\mathbb{R}, +)$, but the subgroup of its positive elements $\mathbb{R}_+^\times$ is isomorphic to $(\mathbb{R}, +)$. Prove this. Hint: The exponential map is a homomorphism $(\mathbb{R}, +) \to \mathbb{R}_+^\times$.

7. In a group G denote by $Z(G)$ the things which commute with every element of G. $Z(G)$ is called the *center* of G. Show $Z(G)$ is a normal subgroup of G.

Given a homomorphism $f : G_1 \to G_2$, sometimes it is helpful to write the sequence of subgroup, group, quotient group

$$(1) \to Ker f \to G_1 \to f(G_1) \to (1).$$

Such a sequence is called *exact*.

If G is a group and α is an isomorphism of G to itself, we say that α is an *automorphism* of G. The set of all automorphisms of G will be denoted by $Aut(G)$. Clearly, this is a group under composition (it is a subset of $S(G)$). A special class of automorphisms can be defined as follows: To each $g \in G$ define α_g by $\alpha_g(h) = ghg^{-1}$. It is easy to see that $\alpha_g \in Aut(G)$ for all $g \in G$. Such automorphisms are called *inner* automorphisms and their totality is denoted by $I(G)$. Check that $I(G)$ is a subgroup of $Aut(G)$ and that the map $g \mapsto \alpha_g$ is a homomorphism whose kernel is $Z(G)$. Hence, $I(G)$ is isomorphic with $G/Z(G)$.

We now turn to a notion which will be very useful in our study of geometry, namely, that of a *group action*. Let G be a group, X be a set and $\phi : G \times X \to X$ be a mapping, called the action of G on X. Writing $\phi(g, x)$ as $g \cdot x$, we shall assume that an action (G, X) satisfies

1. For all $x \in X$, $1 \cdot x = x$

2. For all $g, h \in G$ and $x \in X$, $(gh) \cdot x = (g \cdot)(h \cdot x)$

When 1. and 2. are satisfied we shall call X a G-space or equivalently, say that G acts as a *transformation group* on X. In our applications X will usually be some geometric space and G will be a group of transformations of X which preserves some geometric property such as length, or angle, or area, etc. As we shall see such things always form a group and it is precisely the properties of this group which is the key to understanding length, or angle, or area, respectively. This is the essential idea of Klein's Erlangen Program.

Let (G, X) be a group action. For $x \in X$ we define the *G-orbit* of x, written $\mathcal{O}_G(x)$ as $\{g \cdot x : g \in G\}$. By taking $g = 1$ we see that for each $x \in X$, $x \in \mathcal{O}_G(x)$, so that each orbit is non-empty. Also if $\mathcal{O}_G(x) \cap \mathcal{O}_G(y) \neq \phi$, then $\mathcal{O}_G(x) = \mathcal{O}_G(y)$. For suppose $g_1 \cdot x = g_2 \cdot y$. Then $g_2^{-1} g_1 \cdot x = y$ so $g_3 g_2^{-1} g_1 \cdot x = g_3 \cdot y$. Thus $\mathcal{O}_G(y) \subset \mathcal{O}_G(x)$. Similarly, $\mathcal{O}_G(x) \subseteq \mathcal{O}_G(y)$ so $\mathcal{O}_G(x) = \mathcal{O}_G(y)$. Thus X is the disjoint union of its orbits and we call this the decomposition of X into its G-orbits. In particular if X is finite, then $|X| = \sum_{x \in X} |\mathcal{O}_G(x)|$, where $|S|$ stands for the number of elements of the set S.

Given a group G, an interesting example of an action is provided by taking $I(G)$ for the "group", G for the "set" and taking as the action the natural action of $I(G)$ on G. Then the $I(G)$-orbit of an $h \in G$ is $\{ghg^{-1} : g \in G\}$, called the *conjugacy class* of h, the elements of the center being the ones with trivial, i.e., one point conjugacy classes.

A subset Y of X is called *G-invariant* if $G \cdot Y \subset Y$. When we have an invariant set we get a new action of G on Y. Similarly, if (G, X) is a group action and H is a subgroup of G then (H, X) is also a group action.

Exercise 2.1.4.

1. An important example of a group action is furnished by a group G acting on itself by *right translation*. Prove this is an action.

2. More generally, let G be a group and H be a subgroup. Consider the action of H on G by right translation. Show that its decomposition into orbits gives the set G/H of left cosets.

We shall say that an action (G, X) is *transitive* if for any two elements x and $y \in X$ there is some $g \in G$ so that $g \cdot x = y$. In other words, there is only one orbit. Just as for groups we must now also decide when two group actions (G, X) and (G, Y) are essentially the same, i.e., except perhaps for changes in the names. We shall say that (G, X) and (G, Y) are *G-equivariant* if there is map $f : X \to Y$ which for all $g \in G$ and $x \in X$ satisfies

$$f(g \cdot x) = g \cdot f(x).$$

We call such an f a G-equivariant isomorphism or equivalence, if, in addition, f is bijective.

Let G be a group, H be a subgroup, and let G act on G/H, the set of left cosets, by left translation. This is clearly a transitive action. The following theorem shows that essentially it is the only one.

Theorem 2.1.5. *Let (G, X) be an action and $x \in X$ be fixed. Then the orbit $\mathcal{O}_G(x)$ is a G-invariant set; in fact it is the smallest G-invariant set containing x. Hence this gives a transitive action $(G, \mathcal{O}_G(x))$. More generally, let (G, X) be any transitive action and $x \in X$. Then the stabilizer,*

$$Stab_G(x) = \{g \in G : g \cdot x = x\}$$

is a subgroup of G and the action of G on $G/Stab_G(x)$ by left translation is a G-equivariant equivalence with the action of G on the orbit $\mathcal{O}_G(x)$.

Proof. The fact that $\mathcal{O}_G(x)$ is G-invariant and G acts transitively on it is immediate. The same may be said as far as the stabilizer being a subgroup of G. The only thing that remains to be proven is that we have an equivariant equivalence of actions. To see this call the map $f : G \to X$ which takes $g \mapsto g \cdot x$. Then f is onto and induces a bijection $f^{\sim} : G/Stab_G(x) \to X$, which is easily seen to be G-equivariant. □

Exercise 2.1.6.

1. In the situation of the theorem above let y be another element of X. By transitivity choose $g \in G$ so that $g \cdot x = y$. Show that $gStab_G(x)g^{-1} = Stab_G(y)$. When this happens we say these subgroups are *conjugate*.

2. Observe that conjugate subgroups of a group are isomorphic.

3. Let $f : G_1 \to G_2$ be a group homomorphism and let G_1 act on $f(G_1)$ by left multiplication through f. Show this is a transitive action. Show that the stability group is $Ker f$. Show that the G_1-equivariant equivalence given in the theorem above of this action with the action of G_1 on the orbit $f(G_1)$ gives the first isomorphism theorem.

2.2 Applications to Finite Fields and the Theory of Numbers

Here we discuss some important facts concerning the integers, which lead to a proof of the theorem of Fermat mentioned earlier. Now the integers have two operations $+$ and $\cdot$. They form an Abelian group with respect to the former and are closed, associative and have an identity with respect to the latter, but most elements don't have integer inverses. (It was precisely for this reason that we constructed the rational numbers). The same statements apply equally well to the Gaussian integers and polynomials with coefficients from a field. As mentioned earlier, these as well as many other examples can be unified under the notion of a *commutative ring with identity* R. This is a set with two operations $(R, +, \cdot)$ satisfying the conditions mentioned just above. (Of course, a field is also a commutative ring with identity). We can treat R in a way similar to that of a group. In particular, there are notions of *subring*, *quotient ring* and *ring homomorphism*, the latter being a map from one ring to another which preserves both operations. A subring S of R is called an *ideal* in R if $r \cdot s \in S$ whenever $r \in R$ and $s \in S$. As an exercise the reader should show, exactly as in the case of a group, that:

1. If S is an ideal in R, a commutative ring with identity, then the quotient ring R/S is also a commutative ring with identity.

2. S is an ideal in R if and only if $S = Ker f$, where f is a ring homomorphism.

We recall section 1.8. Now let $n \geq 1$ be a fixed integer and consider $\mathbb{Z}_n = \mathbb{Z}/n\mathbb{Z}$, where $n\mathbb{Z}$ means $\{kn : k \in \mathbb{Z}\}$. Since this is an ideal and $\mathbb{Z}$ is a commutative ring with identity, the same is true of $\mathbb{Z}_n$, which is called the ring of integers modulo n. If $\pi : \mathbb{Z} \to \mathbb{Z}_n$ is the natural ring homomorphism then clearly, for integers x and y, $\pi(x) = \pi(y)$ if and only if $x - y$ is divisible by n, or $x \equiv y (\bmod n)$. Sometimes we will even suppress the n and just write $x \equiv y$. But we will always use the congruence symbol to distinguish it from equality of integers. We shall sometimes write $\pi(x) = \bar{x}$. For example, $\pi(n) = \bar{0}$.

Theorem 2.2.1. *$\mathbb{Z}_n$ is a field if and only if n is a prime.*

As a corollary we get the so called "little" Fermat theorem.

Corollary 2.2.2. *Let p be a prime. Then $a^p \equiv a(\bmod p)$ for every $a \in \mathbb{Z}$.*

Proof. If a is divisible by p, say $a = kp$, then $a^p - a = k^p p^p - kp$ which is divisible by p. Therefore we may assume that a is not congruent to $0(\bmod p)$. Now the multiplicative group $\mathbb{Z}_p^\times$ is a group of order $p-1$ which contains $\bar{a}$. Hence, by Lagrange's theorem $(\bar{a})^{p-1} = \bar{1}$. Unravelling, this tells us $a^{p-1} \equiv 1(\bmod p)$. Finally, multiplying by a gives $a^p \equiv a(\bmod p)$. □

Corollary 2.2.3. *If p is a prime, then the multiplicative group, $\mathbb{Z}_p^\times$, is cyclic.*

Proof. By Fermat's little theorem, each $a \in \mathbb{Z}$ satisfies $a^p \equiv a(\bmod p)$. That is, in $\mathbb{Z}_p$ each element satisfies $x^p - x = 0$. Since there are precisely p elements in $\mathbb{Z}_p$ and this equation has at most p roots, it must have exactly p simple roots (bo repeated). Thus, we get a factorization over $\mathbb{Z}_p$: $x^p - x = (x - \alpha_1)\ldots(x - \alpha_p)$, where the $\alpha_j \in \mathbb{Z}_p$. Now one of these, say $\alpha_p = 0$. Dividing in the polynomial ring gives $x^{p-1} - 1 = \Pi_{\alpha \neq 0 \in \mathbb{Z}_p}(x - \alpha)$. Thus the elements of $\mathbb{Z}_p^\times$ are the $p-1$ roots of unity and therefore $\mathbb{Z}_p^\times$ is cyclic. □

Corollary 2.2.4. *(Wilson's theorem)* $(p-1)! \equiv (-1)(\bmod p)$, *for each prime p.*

Proof. For $p = 2$ this is evidently true. We can therefore assume that p is an odd prime. By Fermat's little theorem, for each $\bar{a} \neq 0 \in \mathbb{Z}_p$ we have $(\bar{a})^{p-1} = \bar{1}$. Consider the polynomial equation $x^{p-1} - \bar{1} = 0$, where this polynomial has coefficients in the field $\mathbb{Z}_p$. Since this is a polynomial of degree $p-1$ and every element of $\mathbb{Z}_p^\times$, a set of order $p-1$ satisfies this equation, they comprise *all* the roots. Hence $x^{p-1} - \bar{1}$ factors into a product of simple roots $(x - \bar{1})(x - \bar{2})\ldots(x - \overline{(p-1)})$. Evaluating at 0 then gives

$$-\bar{1} = (-\bar{1})(-\bar{2})\ldots(-\overline{(p-1)}) = \bar{1}\bar{2}\ldots\overline{(p-1)(-1)}^{p-1}.$$

But since p is odd, $p-1$ is even. Therefore $(-1)^{p-1} = 1$ and $-\bar{1} = \bar{1}\bar{2}\ldots\overline{(p-1)}$. Unravelling, we see that $(p-1)! \equiv (-1)(\bmod\, p)$. □

We are now in a position to prove Fermat's theorem concerning primes which are sums of two squares. The remainder of this section is devoted to this goal and is optional. We first need two lemmas.

Lemma 2.2.5. *Suppose p is a prime and c is a relatively prime integer, i.e., $(p,c) = 1$. If $pc = x^2 + y^2$ is a sum of two squares, then p is also a sum of two squares.*

Proof. By assumption and what we proved above, pc is not prime in $\mathcal{G}$, the Gaussian integers. We want to show the same is true of p. Suppose not. Then, by our results on $\mathcal{G}$, we may assume that p is a prime in $\mathcal{G}$. Since p clearly divides $pc = (x+iy)(x-iy)$ (even in $\mathbb{Z}$!), it must divide either $x+iy$ or $x-iy$. Suppose it divides the former (the reasoning in the other case is identical). Then $x+iy = (a+ib)p$. Hence $x = ap$ and $y = bp$. But then p also divides $x - iy$. This means that p^2 divides $x^2 + y^2 = pc$. By unique factorization in $\mathbb{Z}$, p divides c, a contradiction. □

Lemma 2.2.6. *Suppose p is a prime of the form $4n+1$. Then in $\mathbb{Z}$ there is a solution to $x^2 \equiv -1(\bmod\, p)$, where $0 \leq x \leq p-1$.*

Proof. First we show if there is a solution to the congruence at all, there must already be one where $0 \leq x \leq p-1$. Using the Euclidean algorithm write $x = qp + x_0$, where $0 \leq x_0 \leq p-1$ and x solves the congruence. Then $x^2 = q^2p^2 + 2qpx_0 + x_0^2$ so that $x^2 \equiv x_0^2(\bmod\, p)$. We now prove there is a solution to the congruence. Since $p = 4n+1$, $\frac{p-1}{2}$ is an even integer. Let $x = \frac{p-1}{2}!$. Then clearly x also equals $(-1)(-2)\ldots(-\frac{p-1}{2})$. So $x^2 \equiv (1)(2)\ldots(\frac{p-1}{2})(-\frac{p-1}{2})\ldots(2)(1)$. But since, for each $k \in \mathbb{Z}$, $-(\frac{p-k}{2}) \equiv \frac{p+k}{2}(\bmod\, p)$, we see that $x^2 \equiv (1)(2)\ldots(\frac{p-1}{2})(\frac{p+1}{2})(\frac{p+3}{2})\ldots(p-1)$. Thus $x^2 \equiv (p-1)!(\bmod\, p)$. On the other hand, since Wilson's theorem tells us that $(p-1)! \equiv -1(\bmod\, p)$, we see that $x^2 \equiv -1(\bmod\, p)$. □

We now begin the proof of Fermat's theorem. Let p be a prime of the form $p = 4n+1$. By Lemma 2.2.6 there is an $x \in \mathbb{Z}$ satisfying

$x^2 \equiv -1 (\bmod p)$, where $0 \leq x \leq p-1$. Since p divides x^2+1 i.e., $x^2+1 = cp$, by Lemma 2.2.5 we need only see that $(p, c) = 1$ to conclude that p is a sum of squares of two integers. If not, then p divides c. But then p^2 divides x^2+1. In particular, since everything is positive, $p^2 \leq x^2+1$. On the other hand, $x^2+1 \leq (p-1)^2+1 = p^2-2p+2$. Since this means that $p^2 \leq p^2 - 2p + 2$ i.e., that $p \leq 1$ we get a contradiction. Thus p is a sum of two squares. Before proving the converse we need the following

Proposition 2.2.7. *Let p be a prime $p > 2$. Then $\{y^2 : y \in \mathbb{Z}_p^\times\}$ is a subgroup of index 2 in the group $\mathbb{Z}_p^\times$. This subgroup is the kernel of the homomorphism $x \mapsto x^{\frac{p-1}{2}}$ with values in $\{\pm 1\}$. In other words, there is an exact sequence*

$$(1) \to (\mathbb{Z}_p^\times)^2 \to \mathbb{Z}_p^\times \to \{\pm 1\} \to (1).$$

Proof. Since p is odd, $\frac{p-1}{2}$ is an integer and therefore $x \mapsto x^{\frac{p-1}{2}}$ is a homomorphism. Moreover, since $x^{p-1} = (x^{\frac{p-1}{2}})^2 = 1$ for each $x \in \mathbb{Z}_p^\times$, it follows that $x^{\frac{p-1}{2}} = \pm 1$. Let Ω be the algebraic closure of $\mathbb{Z}_p$.[1] For each $x \in \mathbb{Z}_p^\times$, there is always some $y \in \Omega$ such that $y^2 = x$. Then $y^{p-1} = x^{\frac{p-1}{2}}$. But since $x^{p-1} = 1$, $y^{p-1} = \pm 1$. For x to be a square in $\mathbb{Z}_p^\times$ it is necessary and sufficient that $y \in \mathbb{Z}_p^\times$, that is, $y^{p-1} = 1$. (See proof of the fact that $\mathbb{Z}_p^\times$ is cyclic.) Thus, the set of squares of $\mathbb{Z}_p^\times$ is the kernel of the homomorphism $x \mapsto x^{\frac{p-1}{2}}$. □

Turning to the converse of Fermat's theorem, suppose a prime $p > 2 \in \mathbb{Z}$ is a sum of two squares, $p = x^2 + y^2$. We will prove $p = 4n+1$. If not, then $p = 4n+3$, and so $\frac{p-1}{2}$ is odd. However, the congruence $x^2 \equiv -y^2 (\bmod p)$ can have no solutions in $\mathbb{Z}$, yielding a contradiction. This is because $(-y^2)$ is not a square in $\mathbb{Z}_p^\times$. For if it were, then by the proposition, $(-y^2)^{\frac{p-1}{2}} \equiv 1 (\bmod p)$, which is impossible since $\frac{p-1}{2}$ is odd and $p > 2$.

[1] Any field can be imbedded in algebraically closed field. $\mathbb{Q} \subset \mathbb{C}$ is an example of such an imbedding.

2.3 Linear Algebra

Here we gather together the various facts of what is generally called *linear algebra* which will be needed in our study of geometry in the following chapter.

We shall parameterize the Euclidean plane, $\mathbb{R}^2$, by points $X = (x_1, x_2)$, where for each i, the $x_i \in \mathbb{R}$, the real numbers. These are, respectively, called the first and second coordinates of X. By definition they determine and are determined by $X \in \mathbb{R}^2$. The point in the plane both of whose coordinates are 0 is special, is called the *origin* and is denoted by O. If we want to regard X as a *vector*, we draw an arrow at the end of the line joining X to O . Given vectors $X = (x_1, x_2)$ and $Y = (y_1, y_2)$ we add them by adding the respective coordinates. Scalar multiplication by a real number is also coordinatewise.

$$X + Y = (x_1 + y_1, x_2 + y_2) \tag{2.1}$$

$$tX = (tx_1, tx_2) \tag{2.2}$$

Under these operations is easy to check that $(\mathbb{R}^2, +)$ is an Abelian group whose identity is the origin. In addition, scalar multiplication enjoys the following properties.

1. $t(X + Y) = tX + tY$
2. $(s + t)X = sX + tX$
3. $(st)X = s(tX)$ and
4. $1X = X$

When one has an Abelian group, together with a scalar multiplication with these properties, we call this a vector space over a field (in our case $\mathbb{R}$). For example, we can make the same construction in 3-dimensional space $\mathbb{R}^3$, or indeed in a space of any dimension. Of course a space of dimension 1 is essentially (that is, isomorphic with) $\mathbb{R}$.

Let $\mathcal{W}$ stand for such a vector space *over the field* $\mathbb{R}$. Just as for homomorphisms of groups we will have to consider linear transformations of vector spaces. In our case what will be of greatest importance will be linear transformations. These are mappings satisfying

$L(X+Y) = L(X) + L(Y)$ and $L(tX) = tL(X)$ for all X and $Y \in \mathcal{W}$ and $t \in \mathbb{R}$. For example, it is geometrically clear that in $\mathbb{R}^2$ a rotation about the origin through any angle is linear . The same may be said for scalar multiplication by any scalar .

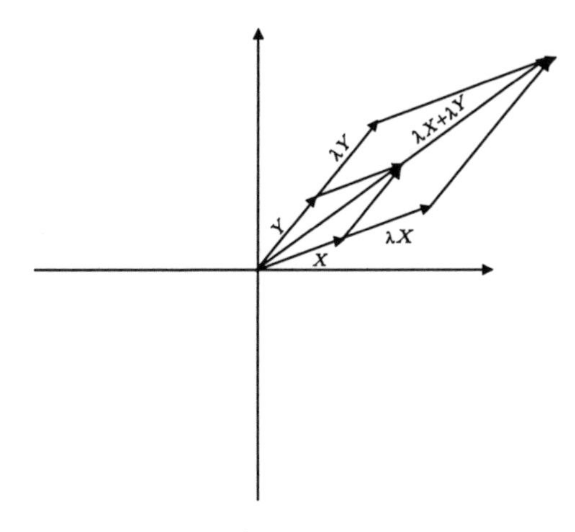

Figure 2.1

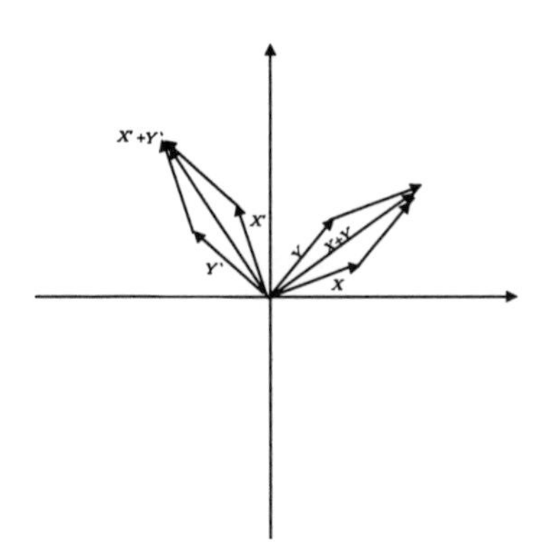

Figure 2.2

A square matrix of order n with coefficients from a field F is an $n \times n$ array of elements from F.

We will now discuss briefly the relationship between linear transformations and matrices. Let $E_1 = (0,1)$ and $E_2 = (1,0) \in \mathcal{W}$. Then any $X = (x_1, x_2) \in \mathcal{W}$ can be written $X = x_1E_1 + x_2E_2$. Such an expression is called a *linear combination* of E_1 and E_2. Clearly, this way of writing X is unique in so far as the choice of coefficients is concerned. Now suppose that L is a linear transformation of $\mathcal{W}$ to iteself. Then $L(X) = x_1 \cdot L(E_1) + x_2 \cdot L(E_2)$ so L is determined by what it does on E_1 and E_2. But, since $L(E_1)$ and $L(E_2) \in \mathcal{W}$, by the above, they in turn can be expressed uniquely as

$$\begin{aligned} L(E_1) &= a_{11}E_1 + a_{12}E_2 \\ L(E_2) &= a_{21}E_1 + a_{22}E_2. \end{aligned} \tag{2.3}$$

Such a 2×2 square array of numbers

$$\begin{pmatrix} a_{11} & a_{12} \\ a_{21} & a_{22} \end{pmatrix}$$

i.e a 2×2 *matrix*. Since it is uniquely determined by L it is called *the matrix of the linear transformation* L. Conversely, of course, any 2×2 matrix gives rise to a unique linear transformation via (2.3). Thus there is a 1:1 correspondence between linear transformations of $\mathcal{W}$ and 2×2 matrices. In an exactly analogous way there is a 1:1 correspondence between linear transformations of a 3 dimensional space and 3×3 matrices.

We now turn to an important feature of this correspondence. Clearly, if L and M are linear transformations of $\mathcal{W}$, so is their composition, LM. Suppose the corresponding matrices are

$$L = \begin{pmatrix} a_{11} & a_{12} \\ a_{21} & a_{22} \end{pmatrix} \tag{2.4}$$

and

$$M = \begin{pmatrix} b_{11} & b_{12} \\ b_{21} & b_{22} \end{pmatrix}. \tag{2.5}$$

Then what is the matrix of LM? For each i, $LM(E_i) = L(M(E_i)) = L(b_{i1}E_1 + b_{i2}E_2)$. By linearity of L this is $b_{i1}L(E_1) + b_{i2}L(E_2)$. Using the matrix forms of L and M above, this is $b_{i1}(a_{11}E_1 + a_{12}E_2) + b_{i2}(a_{21}E_1 +$

$a_{22}E_2) = (a_{11}b_{i1} + a_{21}b_{i2})E_1 + (a_{12}b_{i1} + a_{22}b_{i2})E_2$. We shall take this as the definition of multiplication of the matrices L and M. Thus matrix multiplication corresponds to composing the linear transformations.

Exercise 2.3.1. Taking the obvious definitions of adding matrices by adding the respective entries and scalar multiplication of matrices by a real number also componentwise, show that this corresponds to the operations of adding $L + M$ and scalar multiplying tL.

An important question is the following: which of these linear transformations is *invertible*? These will automatically form a group which is called the general linear group and is usually denoted $\mathrm{GL}(\mathcal{W})$.

Applications: As an example of the use of matrices in helping to organize things let us consider the so called *affine transformations* $A_1(\mathbb{R})$ of the real line (1-dimensional space). These consist of the group generated by dilations[2] by positive real numbers, reflections[3] and translations [4]. Now, the inverse of a dilation, reflection or translation is again a thing of the same type so we don't have to worry about inverses. Nevertheless, since we can apply a finite number of these affine motions, in any order, picturing which elements this group actually consists of can be a little daunting. First notice that reflections through 0 can be absorbed if we enlarge the dilations to dilations by any real number a. Composing these with an appropriate translation would then give reflections about the other points. Now the translations form a subgroup isomorphic with the additive group of $\mathbb{R}$, which is easily seen to be normal in $A_1(\mathbb{R})$. This means that each affine transformation is a product of a dilation (by $a \in \mathbb{R}^\times$) with a translation (by $b \in \mathbb{R}$).

$$\alpha(x) = ax + b, x \in \mathbb{R}$$

We now associate this α with the matrix

$$A = \begin{pmatrix} a & b \\ 0 & 1 \end{pmatrix}. \tag{2.6}$$

[2] A dilation means stretching or shrinking a vector by a positive number.

[3] Replacing the vector by its negative.

[4] Adding a fixed vector to everything undergoing the translation.

A quick calculation shows that multiplication of such matrices corresponds exactly to composing the corresponding affine transformations. Thus the group $A_1(\mathbb{R})$ is isomorphic with the group of matrices described in (2.6). Notice that all these matrices are invertible.

We now come to the important notion of an *inner product* on $\mathcal{W}$. This is a map $\langle , \rangle : \mathcal{W} \times \mathcal{W} \to \mathbb{R}$ satisfying

1. $\langle X, Y \rangle = \langle Y, X \rangle$ (symmetry)
2. $\langle X + Y, Z \rangle = \langle X, Z \rangle + \langle Y, Z \rangle$ and $\langle tX, Y \rangle = t \langle X, Y \rangle$ (linearity)
3. $\langle X, X \rangle \geq 0$ and $\langle X, X \rangle = 0$ only if $X = O$ (positive definiteness)

We leave it to the reader to verify that in $\mathbb{R}^2$ such an inner product is given by

$$\langle X, Y \rangle = x_1 y_1 + x_2 y_2.$$

The main point is that a sum of squares of real numbers is greater than or equal to 0 and can equal 0 only if each of them is zero. Similarly, if we were interested in $\mathbb{R}^3$ we would take for our inner product $\langle X, Y \rangle = x_1 y_1 + x_2 y_2 + x_3 y_3$.

Just as we were interested in which transformations were linear (that is, preserved the linear structure), we now consider which of these preserve the inner product. These will turn out to be automatically invertible and so will form a subgroup called the orthogonal group denoted $O(\mathcal{W})$. These notions can be used to classify the conic sections via 2×2 symmetric matrices. But we will not deal with this here.

Chapter 3

Two and Three Dimensional Geometry and Topology

In this section we will introduce our study of 2-dimensional geometry. Following Descartes, our methods will largely be analytic. We will study both Euclidean and the latter two types of 2-dimensional non-Euclidean geometries. These are respectively called Elliptic (or Spherical) Geometry and Hyperbolic Geometry. For these and other reasons it will also be necessary for us to venture out into 3-dimensional Euclidean geometry. As we shall see these other geometries can be *modelled* in Euclidean space and so their consistency is as good as that of Euclidean geometry itself. This is one of the great discoveries of 19th century mathematics and is associated with the names of Bolyai, Lobachevski , Riemann and Gauss. It is hoped that the study of non-Euclidean geometry will also serve to better clarify Euclidean geometry itself.

3.1 The Euclidean Case

As above, in studying the usual Euclidean plane geometry we shall parameterize the Euclidean plane $\mathbb{R}^2 = \mathcal{W}$ by points $X = (x_1, x_2)$, where for each i, $x_i \in \mathbb{R}$. We can also regard $\mathcal{W}$ as an inner product space where the inner product is given by

$$\langle X, Y \rangle = x_1 y_1 + x_2 y_2.$$

From such an inner product we form the *norm*, or *length* of a vector $X = (x_1, x_2)$. Since, as we saw, non negative real numbers have (unique) square roots, this is by definition is the length

$$\| X \| = \sqrt{\langle X, X \rangle} = x_1^2 + x_2^2.$$

We have also seen that $\| X \| \geq 0$ and can only be zero if $X = O$ and also that for all $t \in \mathbb{R}$ and $X \in \mathcal{W}$, $\| tX \| = t \| X \|$. Notice that by this last fact we always get a vector of norm 1 when we divide by the norm, $\| \frac{X}{\|X\|} \| = 1$, for $\|X\| \neq 0$. This process is called normalizing the vector. Finally we define the distance between points X and Y by

$$d(X, Y) = \| X - Y \| .$$

We shall study its properties shortly.

Exercise 3.1.1. Show that for any two vectors X and Y we have

$$2(\| X \|^2 + \| Y \|^2) = \| X - Y \|^2 + \| X + Y \|^2 .$$

Can you think of a geometric interpretation of this?

Suppose we are given two vectors X and Y. Then the properties of an inner product tell us that

$$\langle X - Y, X - Y \rangle = \langle X, X \rangle - 2 \langle X, Y \rangle + \langle Y, Y \rangle . \tag{3.1}$$

Hence,

$$\langle X, Y \rangle = \frac{\| X \|^2 + \| Y \|^2 - \| X - Y \|^2}{2}. \tag{3.2}$$

It will be helpful in understanding the geometry to know that rotations preserve the inner product. Since rotations are linear it follows from (3.2) that all we need is to show that they preserve the norm. But clearly, $d(X, O) = \| X \|$ and since rotations certainly preserve distances they must also preserve norms. Hence

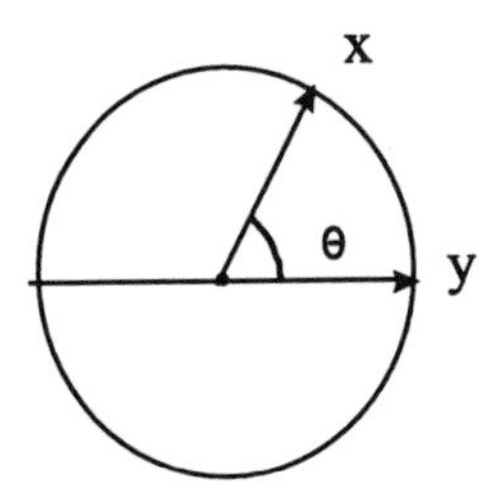

Figure 3.1

Proposition 3.1.2. *A rotation about the origin in* $\mathbb{R}^2$ *preserves the inner product.*

Now what we want to do is understand the relationship between the inner product of two non-zero vectors and the angle θ between them. Here, θ is between 0 and 180 degrees, or, as is more convenient for us, using radian measure, $0 \leq \theta \leq \pi$. Clearly, the angle will not be changed if we normalize the vectors. Let X and Y be two vectors of length 1 and suppose the angle between them is θ. Rotate them so that Y goes to $(1,0)$. Then the angle between the rotated $X = (u_1, u_2)$ and $(1,0)$ is also θ. Since this rotation preserves the inner product $\langle X, Y \rangle = u_1 \cdot 1 + u_2 \cdot 0 = u_1$. Since $\| X \| = 1$, it follows that $u_1 = \cos\theta$. Thus we have proven that the inner product of two vectors of length 1 is the cosine of the angle between them. Now let X and Y be two non-zero vectors and θ be the angle between them. Using the linearity properties of the inner product, we get $\left\langle \frac{X}{\|X\|}, \frac{Y}{\|Y\|} \right\rangle = \frac{\langle X,Y \rangle}{\|X\| \cdot \|Y\|} = \cos\theta$.

Theorem 3.1.3. *If X and Y are non-zero vectors and θ is the angle between them, then $\langle X, Y \rangle = \| X \| \| Y \| \cos\theta$. In particular, X and Y are perpendicular if and only if $\langle X, Y \rangle = 0$.*

Because this theorem is about pairs of vectors and the angle between them, even if we were in 3 space, since all this takes place in a plane (the one spanned by X and Y), the result must *a fortiori* also be true in 3 space! Now the cosine takes its values between -1 and 1 and since the following result certainly holds if either X or $Y = O$, we have

Corollary 3.1.4. *(Cauchy-Schwartz Inequality)* $| \langle X, Y \rangle | \leq \| X \| \| Y \|$

Reasoning similar to the above tells us that the Cauchy-Schwartz Inequality also holds in 3 space. As an exercise it might be interesting to find out when the Cauchy-Schwartz Inequality become an equality?

Corollary 3.1.5. *(Law of Cosines) In any triangle in the plane with sides of length a, b and c, and θ the angle opposite side c, we have*

$$c^2 = a^2 + b^2 - 2ab\cos\theta.$$

Proof. Translate the triangle to the origin. This preserves all distances and angles. Then choose vectors X and Y of appropriate lengths so that X coincides with a and Y with b. By the properties of vector addition $c = \| X - Y \|$. Hence from 3.1 and the theorem we see that $c^2 = a^2 + b^2 - 2ab\cos\theta$. □

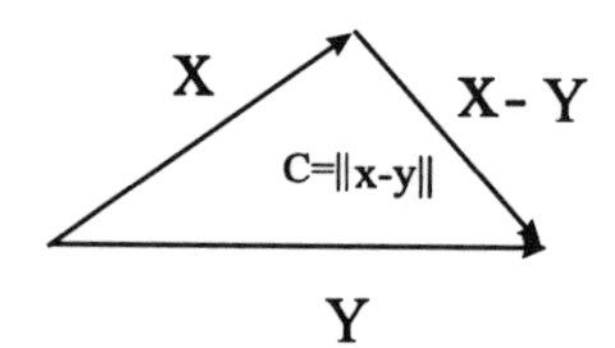

Figure 3.2

Corollary 3.1.6. *In any planar triangle the sides determine the angles. In particular, two triangles with corresponding sides equal are congruent.*

Proof. By the law of cosines, the angle opposite side c is given by $\cos\theta = \frac{a^2+b^2-c^2}{2ab}$. Hence $\cos\theta$ and therefore also θ is determined. The latter follows from the fact that for $0 \leq \theta \leq \pi$ distinct angles have distinct cosines. Since, in this way, each of the angles of the triangle is determined, the result follows. □

Corollary 3.1.7. *(Pythagorean Theorem) Given a triangle in the plane with sides of length a, b and c, where $c \geq a, b$, we have $c^2 = a^2 + b^2$ if and only if the triangle is a right triangle.*

Proof. By the law of cosines $c^2 = a^2 + b^2 - 2ab\cos\theta$. If the triangle is a right triangle, then $\theta = \frac{\pi}{2}$ so $\cos\theta = 0$ and so $c^2 = a^2 + b^2$. Conversely, if $c^2 = a^2 + b^2$, then from the law of cosines $2ab\cos\theta = 0$. Since we are in a field and 2, a, and b are non-zero, we must have $\cos\theta = 0$, so $\theta = \frac{\pi}{2}$ and we are in a right triangle.

□

There are many direct proofs of the The Pythagorean Theorem itself. Here is a particularly neat one.

Let the right triangle have sides a, b, and hypotenuse c. Draw a $c \times c$ square onto side c of the triangle and then draw vertical and horizontal lines through the vertices of this square. It's not difficult to see that this results in three more triangles congruent to the original one. Computing areas we have $(a+b)^2 = c^2 + 4(\frac{1}{2}ab)$. Hence $a^2 + 2ab + b^2 = c^2 + 2ab$ so $a^2 + b^2 = c^2$.

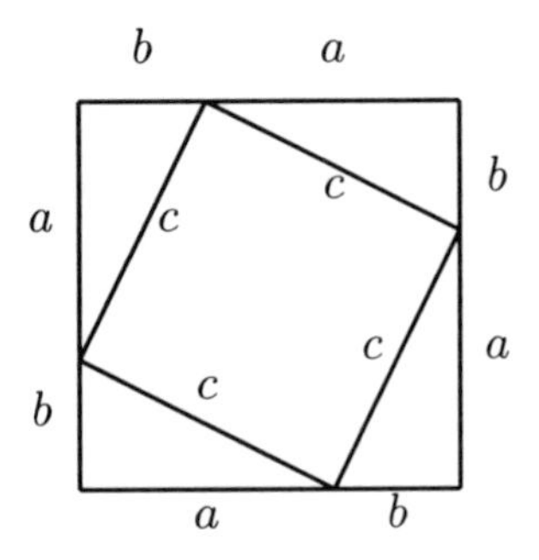

Figure 3.3

Corollary 3.1.8. *The distance between points $d(X,Y)$ satisfies the following properties*

1. $d(X,Y) = d(Y,X)$

2. $d(X,Y) \geq 0$ *and equals* 0 *if and only if* $X = Y$

3. If X, Y *and* Z *are any three points, then* $d(X,Y) \leq d(X,Z) + d(Z,Y)$ *(this last item is also called the triangle inequality)*

Proof. The first two items are not difficult and are left to the reader. We prove the last. It will be sufficient to show that for any two vectors X and Y

$$\| X+Y \| \leq \| X \| + \| Y \| . \tag{3.3}$$

For if X, Y and Z are any three points, then since $X - Y = (X - Z) + (Z - Y)$ we see that $\| X - Y \| \leq \| X - Z \| + \| Z - Y \|$. That is, $d(X,Y) \leq d(X,Z) + d(Z,Y)$. We now turn to the proof of (3.3). Notice that in dimension 1 we have already proven this when we discussed the real numbers. In section 1 we saw that in proving an inequality between positive real numbers we may square both sides and prove that inequality instead. Thus it is sufficient to show that $\langle X+Y, X+Y \rangle \leq (\| X \| + \| Y \|)^2$. Or, $2\langle X,Y \rangle \leq 2 \| X \| \| Y \|$, which follows from the Cauchy-Schwartz Inequality. □

Corollary 3.1.9. *At a point* C *let two directions be given which intersect at some angle, say* θ. *Suppose one goes out a distance* t *from* C *in each direction getting points* $A(t)$ *and* $B(t)$. *Then* $d(t) = d(A(t), B(t))$ *is a linear function of* t.

Proof. By the law of cosines $d(t)^2 = 2t^2(1 - \cos\theta)$. Taking square roots and calling the constant $2(1 - \cos\theta) = k^2$ we see that $d(t) = kt$. □

We shall see that in the hyperbolic and elliptic cases the answer to the analogous question, properly understood, is quite different and this can be said to be the essential difference in character between the three geometries. Later we shall give several other formulations of this *essential difference*.

We now study lines. Given a vector X we get a line $\mathfrak{L}$ through the origin by simply taking all its scalar multiples $\{tX : t \in \mathbb{R}\}$. Thus, a line through the origin is parameterized by a single real parameter t and is determined by a *direction*, namely X. We pick out the origin when $t = 0$. If we are interested in lines $\mathfrak{L}$ which don't necessarily pass through the origin then we simply translate a line through the origin by a fixed vector X_0 .

$$\mathfrak{L} = \{X_0 + tX : t \in \mathbb{R}\}. \tag{3.4}$$

Here again $\mathfrak{L}$ is parameterized by a single real parameter t and is determined by the direction X, but when $t = 0$ we get X_0. Of course, a different X_0 and X can give the same line. Instead of writing the equation as $\phi(t) = X_0 + tX$ we can also write

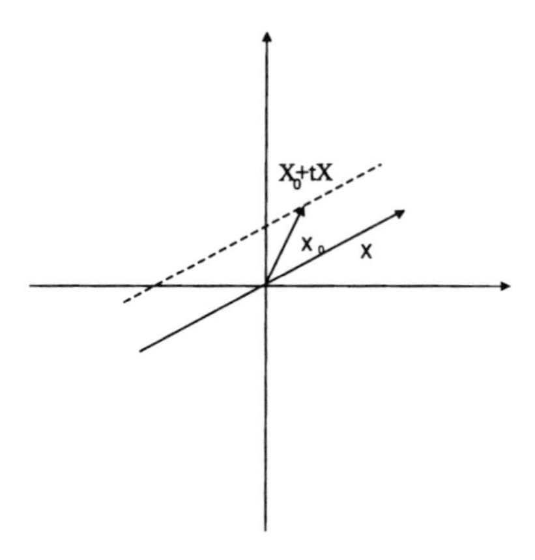

Figure 3.4

$$\phi(t) = (1-t)X_0 + t(X - X_0). \tag{3.5}$$

Then $\phi(0) = X_0$ and $\phi(1) = t(X - X_0)$. For $0 < t < 1$, $\phi(t)$ is on $\mathfrak{L}$ between X_0 and $X - X_0$. The midpoint of the segment joining X_0 and $X - X_0$ occurs when $t = \frac{1}{2}$. (3.5) shows us that each line $\mathfrak{L}$ is *determined* by two points, namely X_0 and $X - X_0$. However, if we want a formula for the distance between points on $\mathfrak{L}$, it is more convenient to use (3.5), which tells us $d(\phi(t), \phi(s)) = |t - s| \parallel X \parallel$. The reader will notice that all that has been said so far about lines works equally well in $\mathbb{R}^3$.

We now turn to the equation of a line. Given $X = (x_1, x_2) \neq O$, we want to find a vector of length 1 perpendicular to X. Clearly, by the theorem above, $\nu = \pm\frac{(-x_2, x_1)}{\|X\|}$ does this and we arbitrarily take the plus sign. Now suppose $\phi(t) = X_0 + tX$ is a line where $X \neq O$, i.e., $\phi(t)$ is not a constant, or to put it another way, the line doesn't degenerate to a point. If Y lies on the line, then

$$\langle Y - X_0, \nu \rangle = 0. \tag{3.6}$$

Conversely, if (3.6) holds, then Y lies on the line. Thus, (3.6) is the equation of the line $\mathfrak{L}$ given by (3.4), where ν is the unit normal vector to the line. If $\nu = (\nu_1, \nu_2)$, (3.6) says that $\nu_1 y_1 + \nu_2 y_2 - \langle X_0, \nu \rangle = 0$, giving the more conventional equation $ay_1 + by_2 + c = 0$.

Corollary 3.1.10. *Any two distinct lines intersect in a unique point unless they have the same normal. In this latter case they have no common points and are called* parallel.

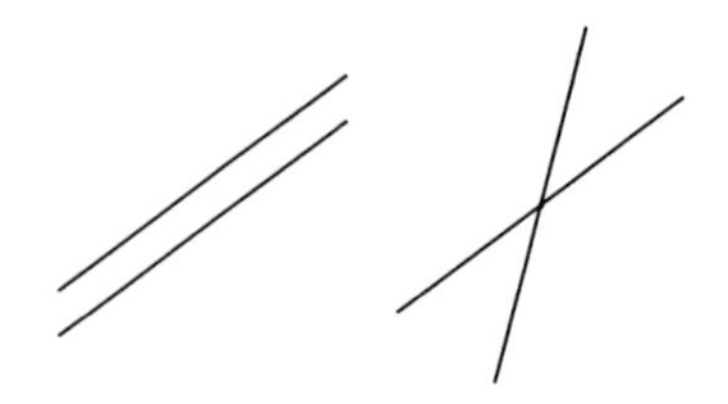

Figure 3.5

This follows from our work on solving two linear equations in two unknowns over a field, in this case $\mathbb{R}$. There is a unique solution except when the normals are proportional, but since they have been normalized and the plus sign chosen, this means they are equal.

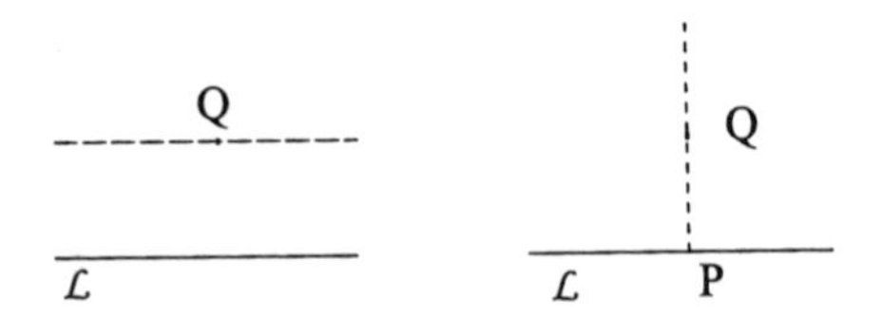

Figure 3.6

Corollary 3.1.11. *Given a line $\mathfrak{L}$ and a point Q not on it, there is a unique line through Q parallel to $\mathfrak{L}$. There is also a unique line* *through Q perpendicular to $\mathfrak{L}$. The distance from Q to $\mathfrak{L}$ is given by $d(Q,\mathfrak{L}) = |\langle Q - Y, \nu\rangle|$, where Y is any point on $\mathfrak{L}$. In particular, $d(Q,\mathfrak{L}) = d(Q,P)$ where P is the unique point of intersection of these perpendicular lines.*

Proof. Let ν be the normal of $\mathfrak{L}$ and take the line through Q with normal ν. The perpendicular line is the one through Q with normal $\pm\frac{X}{\|X\|}$. (Actually, here it doesn't't really matter whether we normalize or take the minus, since we always get the same line.) If P is the unique point of intersection of these perpendicular lines, then by the Pythagorean theorem $d(Q,P)$ is the smallest distance from Q to all points on $\mathfrak{L}$. If Y is any point on $\mathfrak{L}$, then $d(Q,P) = d(Q,Y)\cos\theta$, where θ is the angle between $P-Q$ and $Y-Q$. But since ν is normalized, $\cos\theta = \frac{\langle Q-Y,-\nu\rangle}{\|Q-Y\|}$. Hence, $d(Q,P) = |\langle Q - Y, \nu\rangle|$ □

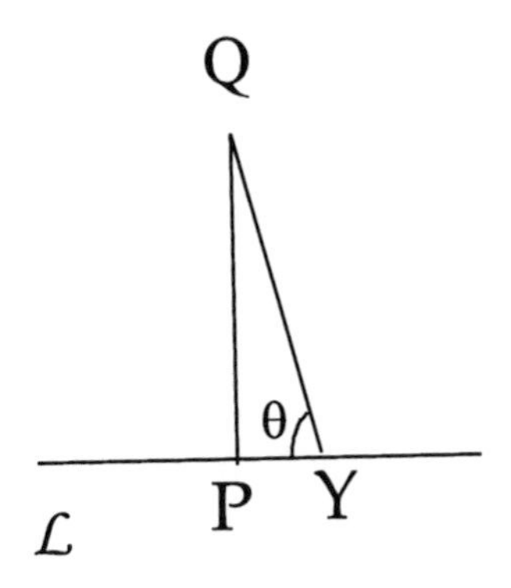

Figure 3.7

The first statement in the result above is called the *Euclidean parallel postulate*. As we shall see, it is what distinguishes Euclidean geometry from that of elliptic and hyperbolic geometry. An equivalent formulation is given in our next result.

Corollary 3.1.12. *In any triangle the sum of the angles is 180 degrees.*

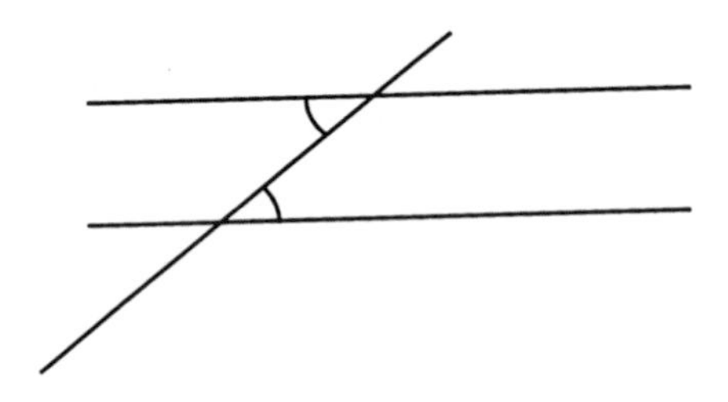

Figure 3.8

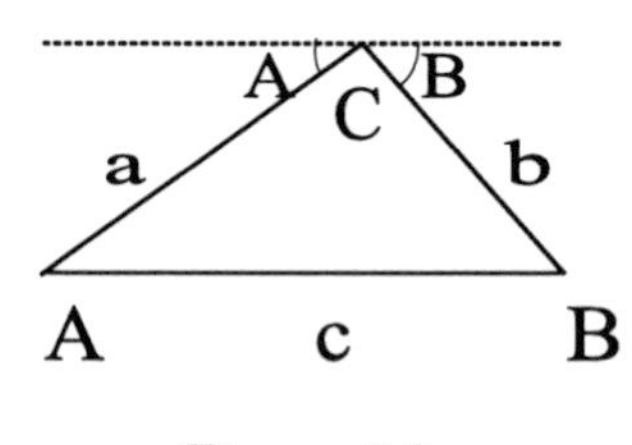

Figure 3.9

Proof. Let the sides be labelled a, b and c, and let A, B and C denote the angles opposite these respective sides, as well as the respective vertices. By the Euclidean parallel postulate, let $\mathfrak{L}$ be the unique line through C parallel to c extended in both directions. The result now follows if we can show that the alternate interior angles formed by pairs of parallel lines are equal . To prove this latter fact translate one of the points of intersection to the origin and erect appropriate perpendiculars . By the corollary immediately above, these have the same length. For the same reason the segments of the parallel lines intercepted by these also have the same length. We conclude from the result following the law of cosines above that the corresponding angles are equal. □

Theorem 3.1.13. *Let $\mathcal{C}$ be a circle and ABC a triangle whose vertices are on the $\mathcal{C}$. Then the angle $\theta = \widehat{CAB}$ is half of the arc $\widehat{BC}$.*

Proof. We have to consider three cases.
Case 1: The center of the circle, O lies on the one of segments AC or CB. For instance let's assume it lies on BC. Then we have:

$$\widehat{BOA} + \widehat{OAB} + \widehat{ABO} = \pi \tag{3.7}$$

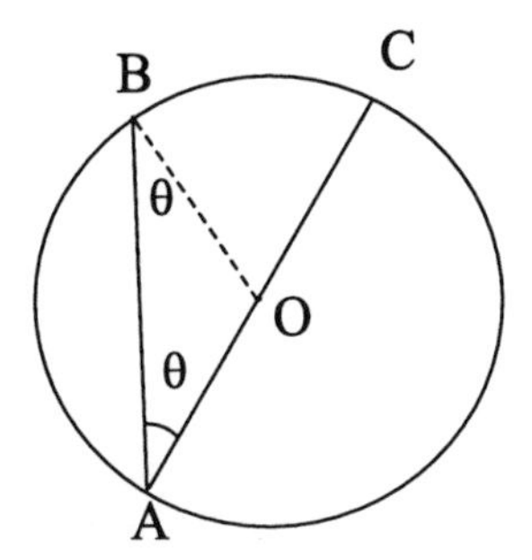

Figure 3.10

We have $\widehat{OAB} = \widehat{ABO}$ and $\widehat{BOA} = \pi - \widehat{BOC}$ and by substituting these in equation 3.7 we get

$$2\widehat{OAB} + \pi - \widehat{BOC} = \pi$$

therefore $\widehat{OAB} = \frac{1}{2}\widehat{BOC}$ and since $\widehat{BOA} = \widehat{CB}$ we conclude that
$\widehat{CAB} = \widehat{OAB} = \frac{1}{2}\widehat{CB}$.

Case 2: The center of the circle, O, is between BA and CA. Connect C to O and continue it till it intersect the arc BC at a point D. By previous case we have:

$$\widehat{BAD} = \frac{1}{2}\widehat{BD}$$

and

$$\widehat{DAC} = \frac{1}{2}\widehat{DC}$$

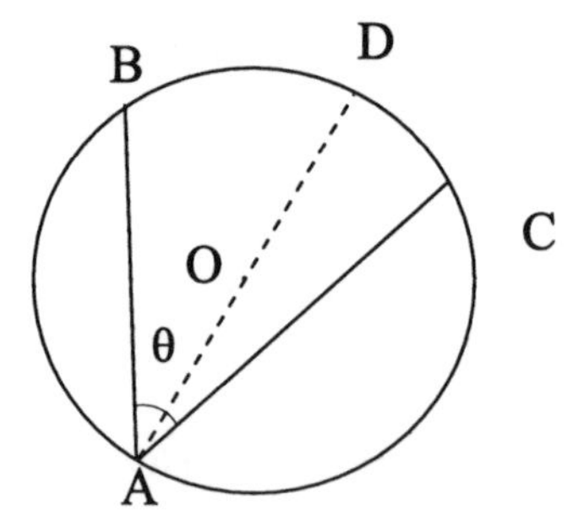

Figure 3.11

by adding these two equations we get

$$\widehat{BAC} = \widehat{BAD} + \widehat{DAC} = \frac{1}{2}\widehat{BD} + \frac{1}{2}\widehat{DC} = \frac{1}{2}\widehat{BC}$$

Case 3: The center of the circle, O, is not between orBA and CA. Again connect C to O by a segment and continue it to intersect the the circle at the point D. Then by the case 1 we have

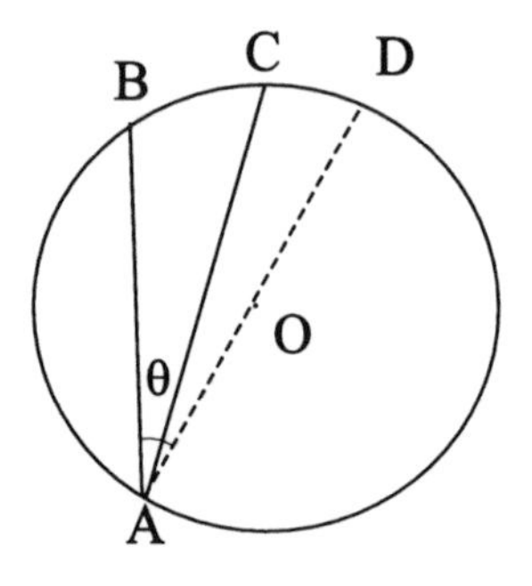

Figure 3.12

$$\widehat{BAD} = \frac{1}{2}\widehat{BD}$$

and

$$\widehat{DAC} = \frac{1}{2}\widehat{DC}.$$

This time by subtracting these two equation we get

$$\widehat{BAC} = \widehat{BAD} - \widehat{DAC} = \frac{1}{2}\widehat{BD} - \frac{1}{2}\widehat{DC} = \frac{1}{2}\widehat{BC}$$

□

Theorem 3.1.14. *Let $\mathcal{C}$ be a circle and BAC a triangle such that B and C are on $\mathcal{C}$ and the A is outside of the circle, then the angle $\widehat{ABC}$ is half of the difference of the two arcs on $\mathcal{C}$ between AB and AC.*

Proof. Let D resp. E be the two points where AB resp. AC meet $\mathcal{C}$. connect D by a segment to C. We have

$$\widehat{DAC} + \widehat{CDA} + \widehat{ACD} = \pi.$$

or

$$\widehat{DAC} = \pi - \widehat{CDA} - \widehat{ACD}.$$

On the other hand $\pi - \widehat{CDA} = \widehat{BDC}$. By substituting this we get

$$\widehat{DAC} = \widehat{BDC} - \widehat{ACD}.$$

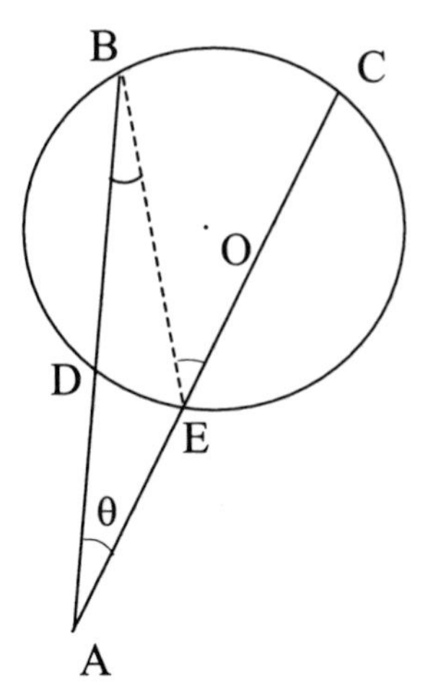

Figure 3.13

But by previous theorem we have

$$\widehat{ACD} = \frac{1}{2}\widehat{DF}, \text{ and } \widehat{BDC} = \frac{1}{2}\widehat{BC}$$

and by substituting these values we get

$$\widehat{DAC} = \frac{1}{2}(\widehat{BC} - \widehat{DF}).$$

□

Theorem 3.1.15. *Let $\mathcal{C}$ be a circle and BAC a triangle such that B and C are on $\mathcal{C}$ and the A is inside the circle, then the angle $\widehat{ABC}$ is half of the sum of the two arcs on $\mathcal{C}$ between AB and AC or their continuation.*

Proof. Let D resp. E be the two points where AB resp. AC meet $\mathcal{C}$. We have

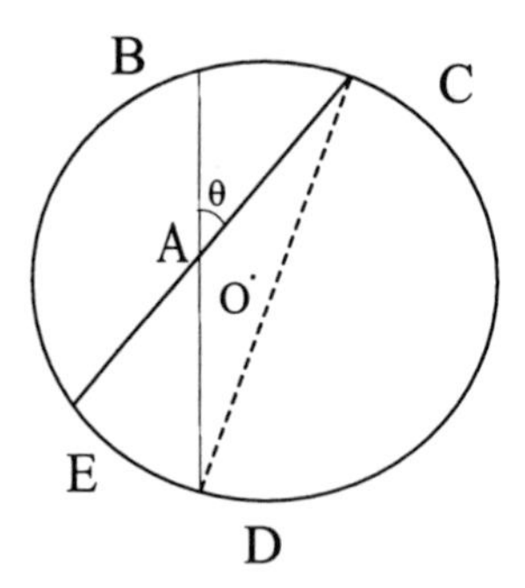

Figure 3.14

$$\widehat{ABE} + \widehat{BEA} + \widehat{EAB} = \pi$$

or

$$\widehat{EAB} = \pi - \widehat{BEA} - \widehat{ABE}$$

and since $\widehat{BEC} = \pi - \widehat{BEA}$ we have

$$\widehat{EAB} = \widehat{BEC} - \widehat{ABE}.$$

By theorem 3.1.13 we have

$$\widehat{BEC} = \frac{1}{2}\widehat{BC} \text{ and } \widehat{ABE} = \frac{1}{2}\widehat{DE}$$

and by substituting the values we get

$$\widehat{CAB} = \frac{1}{2}(\widehat{BC} + \widehat{DE})$$

□

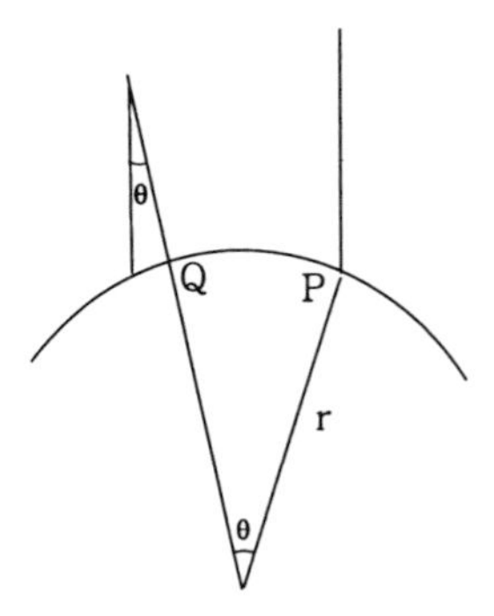

Figure 3.15

As an application of the Parallel Postulate we give an experimental procedure, due to the Greeks, for indirectly measuring the radius of the earth (assuming that one knows the earth is spherical). Imagine two experimenters on opposite sides of a valley several miles wide, in the early morning, with a clear view of one another, at points P and Q, respectively. Suppose it is a sunny day, as is often the case in Greece, and that the two Greeks are equipped with mirrors so that they can signal to one another. The experimenters each place a long straight stake vertically in the ground. They first measure the distance $d(P, Q)$ from P to Q. As we approach high noon the sun becomes more and more vertical. When it is *exactly noon* at P and *the stake has no shadow* the experimenter at P signals with the mirror to the experimenter at Q to make a measurement of the angle θ, in radians say, between the rays of light of the sun and the stake at Q by means of the shadow. Now the rays of light, coming from what is essentially *an infinitely distant source*, must be *parallel*. It follows, by the parallel postulate, that *the angle that the lines meet in, drawn by extending the stakes downward to the center of the earth is also* θ . Since the points are quite close to one another as compared to the circumference of the earth there is essentially no difference between $d(P,Q)$ and the sector of the circumference between P and Q. Thus we have a sector of a circle with central angle θ, arc length, $d(P,Q)$ and unknown radius, r. It follows that $\frac{d(P,Q)}{2\pi r} = \frac{\theta}{2\pi}$ and therefore

$$r = \frac{d(P,Q)}{\theta}.$$

We now turn to the question of self mappings of the space. We shall be particularly interested in *isometries*, *dilations* and *affine motions*. In order to describe these mappings we shall use our correspondence with matrices explained earlier. Recall that $\mathrm{GL}(\mathcal{W})$ is the group of invertible linear transformations of $\mathcal{W}$, $O(\mathcal{W})$ the orthogonal ones, and $\mathrm{SO}(\mathcal{W})$ the rotations. Now the rotation, L_θ, through angle θ has the following matrix:

$$L_\theta = \begin{pmatrix} \cos\theta & -\sin\theta \\ \sin\theta & \cos\theta \end{pmatrix} \tag{3.8}$$

One way to verify this is to recall that the rotation of a point $X = (x_1, x_2)$ by angle θ means multiplying $x_1 + ix_2$ by $\cos\theta + i\sin\theta$ as complex numbers, thus verifying (3.8).

We recall that $\det \begin{pmatrix} a & b \\ c & d \end{pmatrix} = ad - bc$. We leave it as an exercise for the reader to verify that $\det L = \det L^t$ where L^t is the transpose of L. Thus if $L = \begin{pmatrix} a & b \\ c & d \end{pmatrix}$ then $L^t = \begin{pmatrix} a & c \\ b & d \end{pmatrix}$.

Exercise 3.1.16. Prove that:

(1) $\det(LM) = \det L \det M$

(2) $\det \begin{pmatrix} \cos\theta & \sin\theta \\ \sin\theta & \cos\theta \end{pmatrix}$

In particular, $\det L_\theta = 1$. Now, more generally, if L is orthogonal this means that $L \cdot L^t = I$. Hence, taking determinants and using the fact that $\det L = \det L^t$, yields $\det^2 L = 1$. so $\det L = \pm 1$. Since there are clearly orthogonal matrices with $\det L = 1$, namely, rotations and, as we shall see below, also those with $\det L = -1$, namely, reflections, it follows that $\det : O(\mathcal{W}) \to \{\pm 1\}$ is a homomorphism onto a multiplicative group consisting of $\{\pm 1\}$. It follows that its kernel $\mathrm{SO}(\mathcal{W})$ is normal subgroup of index 2 in $O(\mathcal{W})$.

A map $T : \mathcal{W} \to \mathcal{W}$ is called an isometry of $\mathcal{W}$ if for all X and $Y \in \mathcal{W}$, $d(T(X), T(Y)) = d(X, Y)$. Clearly, such a map is a bijection and the identity map is an isometry. Also the inverse of an isometry and the composition of two isometries is again an isometry. Thus we have a subgroup of the group of all permutations of $\mathcal{W}$. This group is called the *isometry group*, $Isom(\mathcal{W})$ of $(\mathcal{W}, d)$. When we say that two geometric figures are *congruent*, we mean that one of them maps onto the other by an isometry. Actually, we have already implicitly used isometries in various arguments given above, e.g., translating or rotating a figure into a more convenient position. Our purpose now is to formalize our understanding of the isometry group. Later when we consider non-Euclidean geometries we shall see that they have quite different isometry groups and that these group theoretic differences somehow reflect the geometry itself. The isometry group as well as the other groups we shall deal with all act as transformation groups on the ambient space $\mathcal{W}$.

There are three obvious types of isometries. These are *translations*, *reflections* through a line passing through the origin, and *rotations* about the origin. We shall see that any isometry is a product of these, or that these elements generate $Isom(\mathcal{W})$. We have already described rotations above. They have only the origin as a fixed point unless the rotation is the identity, i.e., is a rotation through an angle which is an integer multiple of 2π. Now we describe translations. Given a vector $X_0 \in \mathcal{W}$ we define $T_{X_0}(X) = X_0 + X$, $X \in \mathcal{W}$. This is called the translation by X_0. It is an isometry because $d(T_{X_0}(X), T_{X_0}(Y)) = \| X_0 + X - X_0 + Y \| = \| X - Y \| = d(X, Y)$. Of course, $T_{X_0} = I$ if and only if $X_0 = 0$. Hence T_{X_0} is only linear if

$X_0 = 0$. Evidently, the map $X \mapsto T_X$ is an isomorphism of $\mathcal{W}$ with the group of all translations. Thus the translations form an Abelian subgroup of the group of all isometries. Clearly, a translation has no fixed points unless it is the translation by O, in which case every point is a fixed point.

For a line $\mathfrak{L}$ in $\mathcal{W}$ satisfying (3.6), where ν is the normal to $\mathfrak{L}$, we write the analytic definition for the reflection $T_{\mathfrak{L}}$ through $\mathfrak{L}$.

$$R_{\mathfrak{L}}(X) = X - 2\langle X - X_0, \nu\rangle \nu \tag{3.9}$$

Notice that if $X \in \mathfrak{L}$, then since $\langle X - X_0, \nu\rangle = 0$, $R_{\mathfrak{L}}$ leaves X fixed. Conversely, if $R_{\mathfrak{L}}$ leaves an $X \in \mathfrak{L}$ fixed, then X satisfies (3.6) and so lies on $\mathfrak{L}$. Since $\mathfrak{L}$ passes through the origin if and only if $X_0 = O$, the equation of a reflection through a line passing through the origin is given by $R_{\mathfrak{L}}(X) = X - 2\langle X, \nu\rangle \nu$. In particular, it's a linear transformation.

Lemma 3.1.17. *The determinant of a reflection through a line passing through the origin is* -1.

Proof. Clearly a reflection $R_{\mathfrak{L}}$ through a line $\mathfrak{L}$ passing through the origin can be transformed by a rotation L_θ to a reflection R_0 about the x-axis. Since $\det L_\theta = 1$ and $R_{\mathfrak{L}} = L_\theta \cdot R_0$, it follows that $\det R_{\mathfrak{L}} = \det R_0$. Thus it is sufficient to prove the lemma for R_0 alone and since the matrix for R_0 is

$$\begin{pmatrix} 1 & 0 \\ 0 & -1 \end{pmatrix}.$$

This is clear. □

From this we see that

Proposition 3.1.18. *Any* $T \in O(\mathcal{W})$ *is a product of a rotation and a reflection about a line passing through the origin.*

Proof. Since the determinant of a reflection through a line passing through the origin is -1 and the index of $SO(\mathcal{W})$ in $O(\mathcal{W})$ is 2, there is a fixed reflection such that each $L \in O(\mathcal{W})$ is a product of something in $SO(\mathcal{W})$ and this reflection. □

Proposition 3.1.19. *Any reflection is an isometry of order 2.*

Proof. We first show that the reflection $R_{\mathcal{L}}$ is an isometry. Let X and $Y \in \mathcal{W}$. Then $d(R_{\mathcal{L}}(X), R_{\mathcal{L}}(Y)) = \| R_{\mathcal{L}}(X) - R_{\mathcal{L}}(Y) \|$. But, $R_{\mathcal{L}}(X) - R_{\mathcal{L}}(Y) = (X - Y) - 2\langle X - X_0 - Y - X_0, \nu\rangle \nu$. That is, $(X - Y) - 2\langle X - Y, \nu\rangle \nu$. Computing the norm we get

$$\| R_{\mathcal{L}}(X) - R_{\mathcal{L}}(Y) \|^2 = \| X - Y \|^2 - 4\langle X - Y, \nu\rangle^2 + 4\langle X - Y, \nu\rangle^2 .$$

Thus $R_{\mathcal{L}}$ is an isometry. To see that $R_{\mathcal{L}}$ is of order 2, that is $R_{\mathcal{L}}^2 = I$, observe that for $X \in \mathcal{W}$, $R_{\mathcal{L}}^2(X) = R_{\mathcal{L}}(X - 2\langle X - X_0, \nu\rangle \nu)$. But this is

$$X - 2\langle X - X_0, \nu\rangle \nu - 2(\langle X - X_0 - 2\langle X - X_0, \nu\rangle, \nu\rangle)\nu,$$

that is, $X - 2\langle X - X_0, \nu\rangle \nu - 2\langle X - X_0, \nu\rangle \nu + 4\langle X - X_0, \nu\rangle \nu = X$. □

Exercise 3.1.20.

1. Prove any reflection is a conjugate by a translation of a reflection through a line passing through the origin. In particular, each reflection is a product of translations and a reflection through a line passing through the origin.

2. Any translation is a product of two reflections through parallel lines. What is the relationship of the distance between the parallel lines and the length of the vector that determines the translation? What is the direction of the vector that determines the translation?

3. Any rotation is a product of two reflections through lines passing through the origin. What is the relationship of the angle between the lines and the angle of rotation?

4. Using the two exercises immediately above together with corollary (3.1.22) below, show that any isometry is a product of reflections.

5. Which linear transformations of $\mathcal{W}$ preserve area, say of triangles? Suggestion: Write the matrix of the linear transformation and experiment with its effect on figures. Are we really limited to triangles?

The proof of the next theorem is more advanced than the other material presented here and may be omitted on a first reading.

Theorem 3.1.21. *An isometry T which leaves the origin fixed must be linear and therefore in $O(\mathcal{W})$.*

Proof. We have already observed, from the law of cosines, that an isometry preserves angles. Let $\epsilon > 0$ and $X \neq O$. Since the angle between ϵX and X is 0, it follows that the angle between $T(\epsilon X)$ and $T(X)$ is also 0. Hence $T(\epsilon X) = k \cdot T(X)$ where k is some real number (depending perhaps on ϵ and X). Now $\| T(\epsilon X) \| = \| \epsilon X \| = \epsilon \| X \|$, and $\| k \cdot T(X) \| = |k| \cdot \| X \|$. It follows that $|k| = \epsilon$ so $k = \pm\epsilon$. Thus k depends only on ϵ and for each $X \neq O$, $T(\epsilon X) = \pm\epsilon T(X)$. Viewing everything as a (continuous) function of ϵ, by connectedness and continuity, it can not take on only two values. Therefore, for all ϵ, either $T(\epsilon X) = \epsilon T(X)$ or $T(\epsilon X) = -\epsilon T(X)$. Since T is an isometry and therefore continuous, the latter is impossible unless T is identically zero, in which case it is certainly linear.

We shall prove our result under the additional assumption that T is differentiable at O. Since $T(O) = O$, $\lim_{\epsilon \to 0} \frac{1}{\epsilon} T(\epsilon X) = dT(X)$, where dT, the derivative of T at O, is a *linear* transformation on $\mathcal{W}$. Now, as we saw, $T(\epsilon X) = \epsilon T(X)$, for all $\epsilon > 0$. Hence, $\frac{1}{\epsilon} T(\epsilon X) = T(X)$. Thus $T(X)$ is the limiting value and therefore $T(X) = dT(X)$ for every $X \neq O$. Since this obviously also holds for $X = O$, $T = dT$, and T is indeed linear. As a linear isometry T is orthogonal. □

Corollary 3.1.22. *Any isometry is a product of a translation and something in $O(\mathcal{W})$.*

Proof. Let T denote the isometry and $X = T(O)$. Then composing with T_{-X} gives us an isometry leaving O fixed. Hence, by the theorem $T_{-X} \cdot T = U$ is orthogonal. But then, since $T_{-X}^{-1} = T_X$. we see that $T = T_X \cdot U$. □

Corollary 3.1.23. *The translations form a normal subgroup of the group of all isometries.*

Proof. For $X \in \mathcal{W}$, $L \in \mathrm{GL}(\mathcal{W})$ and $Y = L(X)$ it is easy to see that $L \cdot T_X \cdot L^{-1} = T_{L(X)}$. In particular, since this is true for $L \in O(\mathcal{W})$ and the translations form an Abelian group, the result follows from the above. □

The following property is known as 2-point transitivity. It is also shared by the non-Euclidean geometries.

Corollary 3.1.24. *Given two pairs of points X_1 and X_2 and Y_1 and Y_2 in $\mathcal{W}$ such that $d(X_1, X_2) = d(Y_1, Y_2)$, there is always an isometry T such that $T(X_1) = Y_1$ and $T(X_2) = Y_2$. In particular, $Isom(\mathcal{W})$ acts transitively on $\mathcal{W}$.*

Proof. Clearly, we can choose translations which take X_1 and Y_1 to O. Then simply apply an appropriate rotation taking X_2 to Y_2. □

Corollary 3.1.25. *If each of two triangles have two angles and the included side equal to the corresponding angles and included side, then they are congruent.*

Proof. By the above, choose an isometry taking the side of one onto the corresponding side of the other. Since the corresponding angles are also equal and these determine the lines and these lines intersect in a unique point, this completely determines the triangles. Hence they are congruent. □

By a similar argument one can prove the following corollary which is left to the reader as an exercise.

Corollary 3.1.26. *If each of two triangles have two sides and the included angle equal to the corresponding sides and the included angle, then they are congruent.*

We conclude our work on the geometry of lines with a result which illustrates, in yet another way, the essential character of Euclidean space. The answer to the corresponding question in the non-Euclidean cases, of course, will be quite different.

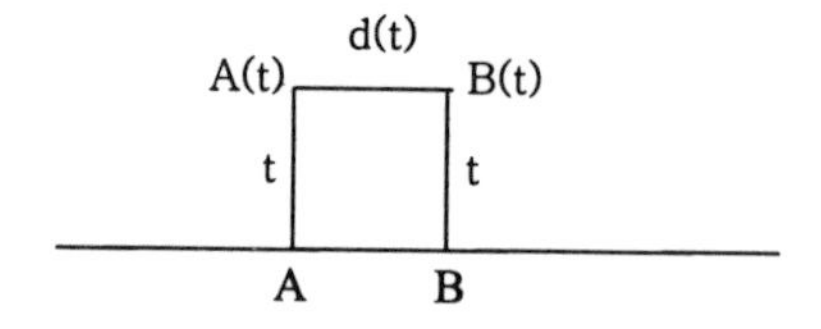

Figure 3.16

Corollary 3.1.27. *Let $\mathfrak{L}$ be a line and A and B be distinct points on $\mathfrak{L}$. Erect perpendicular lines to $\mathfrak{L}$ at these points and go out (in the same direction) a distance t on both of these lines to arrive at $A(t)$ and $B(t)$, respectively, and call $d(t) = d(A(t), B(t))$. Then $d(t) = d(A, B)$ for all t .*

Proof. Draw the line connecting A with $B(t)$. Then angle $A(t)AB(t)$ equals angle $AB(t)B$. Hence the two triangles are congruent since they have two sides and the included angle of one equal to the corresponding two sides and the included angle of the other. It follows that the corresponding sides have the same length; $d(A(t), B(t)) = d(A, B)$. □

Finally, we define an affine motion as a product of a translation, something in $O(\mathcal{W})$ and a scalar multiple of the identity with positive entry, that is, a product of an isometry and a scalar multiple of the identity with positive entry. Thus the affine motion group contains the isometries as a subgroup. As an exercise show that two triangles differ by an affine motion if and only if they are *similar*. This is the significance of the group of affine motions. Alternatively, show that an affine motion which preserves area must be an isometry. (Of course, an isometry preserves area). The dilation group is the linear part of the affine motion group (eliminating the translations). Thus it bears the same relationship to the affine motions as the orthogonal group bears to the isometries. In contrast, we shall see that in non-Euclidean geometry there are no similar figures except for isometric (i.e., congruent) ones. This concludes our study of the geometry of the Euclidean plane. We now turn to non-Euclidean geometry.

3.2 Elliptic and Hyperbolic Geometry

Here we will follow the historical development and begin with hyperbolic geometry, in dimension 2, then turn to 2 dimensional elliptic or spherical geometry, and finally conclude with some remarks on geometry in higher dimensions.

Hyperbolic Geometry:
This is the geometry (for us in 2 dimensional space) in which, given a line $\mathfrak{L}$ and a point P not on $\mathfrak{L}$, there are several distinct lines through P, parallel to $\mathfrak{L}$.

We begin our discussion here with the parallel postulate of Euclid mentioned (and proven) in the previous section. From the time it was formulated in classical antiquity until the 19th century this axiom was considered to be somehow less "self evident" than the others. As in the previous section, the definition of two lines in the plane being parallel means that no matter how far each is extended in either direction they never meet, i.e., have no points in common. Of course, in this definition the kicker, "no matter *how far* extended in either direction" makes this something which can *never be verified by experiment*, since even with a telescope, we can see only a finite distance. By contrast, all the other axioms of Euclid had a finite character and were therefore considered to be more self evident. For this reason many mathematicians over the centuries tried to find a proof of this axiom from the others. it being generally considered a good idea, although not strictly necessary, to have a minimal, or non redundant set of axioms in any mathematical study. This was done particularly in the late 18th and 19th centuries and usually had the character of showing that if the parallel postulate did not hold, then this or that "ridiculous" conclusion would follow. These conclusions, far from being "ridiculous", were in reality the first theorems of non Euclidean geometry, but this was not realized at the time. That was because, during this period, and through the early 19th century, it was considered, particularly by philosophers such as Kant, impermissible to have a system of geometry not in absolute accord with that of Euclid. Of course, until something is proven, one cannot know if failure is merely a lack of diligence or imagination, or both, or whether it is due to some more *fundamental cause*, such as perhaps that what one wishes to prove is false!

In order to address this issue we must define our terms rather more sharply, for this is not exactly an issue of mathematics in the ordinary sense that we have been discussing these matters so far. This is some kind of meta-result, namely, a question of consistency and independence of an axiom system. Now, how can one prove or disprove the statement that the parallel postulate is independent (i.e., cannot be proven from) the other axioms of Euclidean geometry. And indeed how can one prove, if true, that axioms of Euclidean geometry themselves are consistent (i.e don't result in two contradictory theorems somewhere down the road)? It will be necessary for the reader to reflect carefully on these questions and verify, in ones own mind, that these are not questions of mathematics in the usual sense.

The way this was dealt with, more or less simultaneously by Bolyai and Lobachevsky in print and Gauss unpublished, was by constructing a "model" of geometry, that is, something where the primitive terms of "point" and "line" are interpreted in some manner, and under this interpretation all the axioms of Euclidean geometry are "satisfied", *but the parallel postulate is not*. Here, by "satisfied" is meant can be proven, if necessary, by Euclidean geometry (in 2- or perhaps 3-dimensions). Similarly, the parallel postulate not being satisfied must also be proven by Euclidean geometry (but without use of the parallel postulate). The only way we can have such a model, in which Euclidean geometry is dragged into the act, is to have the model sitting inside Euclidean space itself and this is precisely what the above-mentioned 19th century mathematiciana did, and we ourselves will do. Then the consistency of this new geometry will be as "good" as that of Euclidean geometry. It must be mentioned here, however, that the consistency of Euclidean geometry itself remains unproven!

We now turn to a model of Hyperbolic plane geometry. For the "space" we take D, the interior of the unit disk in the plane. The "points" of the space are the ordinary points of D, while the "lines" are arcs of circles which meet the boundary circle of D orthogonally or are ordinary straight lines passing through the center of D (which also meet the boundary orthogonally). In what follows, both in hyperbolic geometry and in elliptic geometry, we shall use the term *geodesic* interchangeably with "straight line". (Remember, if "lines" are regarded as light rays in a physical world model of geometry, Einstein's theory of relativity tells us that light bends in the presence of mass, as well as when it passes through a substance with positive optical density). Continuing a "line" indefinitely means following the circle all the way to, but not including the boundary. Notice that in this model any two distinct points determine a unique line, while any two distinct lines intersect in at most one point. But also observe that there are infinitely many lines passing through a given point (not on some particular line) Later we will explain what the "distance" between points is and see, in particular, that the boundary is infinitely distant. We will also have to understand, just as in Euclidean geometry, what the isometry group is. As opposed to Euclidean geometry, here we shall also have to find other models which reveal new aspects of our geometry, while at the same time making certain features more obscure. In the next section we shall have, to make a pun, a parallel development for the spherical geometry due to Riemann in the mid 19th century. Clearly, all this is a *revolutionary* development and has, in a most significant and conceptual way, affected mathematics and, more generally, intellectual life ever since.

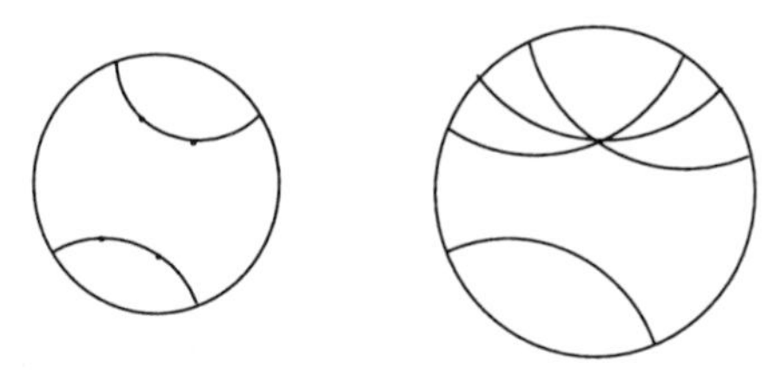

Figure 3.17

Getting to the details, we define distance between the distinct points P and Q in the interior of D as follows. Let γ be the unique geodesic joining the two points. Let A and B be the points where γ intersects the boundary, labelled so that on γ, P is between A and Q and Q is between P and B. Then form the cross ratio,

$$(ABQP) = \frac{\frac{QA}{QB}}{\frac{PA}{PB}}.$$

Then the appropriate definition of the distance between P and Q is given by

$$d(P,Q) = \log(ABQP).$$

Of course, if $P = Q$, then we take $d(P,Q) = 0$. Notice that if P is fixed and Q moves along γ (a circle or a line) toward the boundary, then its hyperbolic distance from P becomes infinite (even though its Euclidean distance from P remains bounded!).

Just as for Euclidean space, we are now interested in understanding the isometry group, i.e those 1:1 and onto mappings of D to itself which preserve hyperbolic distance. This is because the very definition of geometry is the study of those properties of a space which are invariant under the isometry group. (Isometries will automatically also preserve angles and area.) As we shall see below, in our case this will mean the isometries of the interior of D are essentially a conjugate of the group $SL(2,\mathbb{R})$.

Let us consider an invertible 2×2 real, or complex, matrix.

$$T = \begin{pmatrix} a & b \\ c & d \end{pmatrix}.$$

To T we can associate a *linear fractional transformation* of the complex plane, $\mathbb{C}$, as follows: $T'(z) = \frac{az+b}{cz+d}$. Now it is easy to see that T' is a 1:1 and onto mapping of $\mathbb{C}$ to itself. It is easy to check that the map $T \mapsto T'$ is a group homomorphism, whose kernel is the scalar matrices. (Notice if we multiply all the entries of the matrix T by a scalar we don't change the linear fractional transformation T').

We now state, without the proof, the facts about isometries of D and also those which preserve the interior of D.

Theorem 3.2.1. *T' preserves D if and only if $T'(z) = \frac{az+b}{\bar{b}z+\bar{a}}$, $|a| > |b|$. Such T' form a group which acts transitively by isometries. The T''s* *which fix* 0 *are exactly the rotations, R. These are the one with* $b = 0$, *or*

$$T'(z) = \frac{az}{\bar{a}} = \lambda z$$

where $|\lambda| = 1$.

We can therefore apply the theorem on transitive group actions and conclude

$$D = Isom(D)/R.$$

It being understood what a "straight line" is in this situation, we come to the following important result.

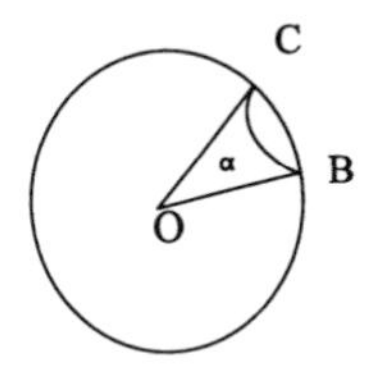

Figure 3.18

Corollary 3.2.2. *The sum of the angles of any triangle is* less *then 180 degrees.*

Proof. Denote the angles of the triangle by A, B, and C. Since the isometry group acts transitively and preserves angles, we can choose an isometry taking A to O. Then the geodesics O, B and O, C are Euclidean straight lines and the hyperbolic angle BOC equals the Euclidean angle BOC. Let γ denote the unique geodesic joining B and C and $\mathfrak{L}$ the Euclidean straight line joining B and C. Then clearly, the hyperbolic angles B and C are less than the respective Euclidean angles B and C. Since in the Euclidean triangle the sum of the angles is 180 degrees, in the hyperbolic triangle this sum must be *less* than 180 degrees. $\square$

It is left to the reader as an exercise to show that this corollary, in turn implies the hyperbolic parallel postulate. Thus the two are actually equivalent.

We conclude our brief discussion of hyperbolic geometry with a statement without proof of the following non trivial fact, due to Gauss. We define the *excess* of a triangle $\mathcal{T}$ to be the amount it differs from 180 degrees, in radians. Thus $ex(\mathcal{T}) = \pi - (A + B + C)$.

Theorem 3.2.3. *For a hyperbolic triangle, its excess equals its area.*

From this we get the following rather curious corollary.

Corollary 3.2.4. *In hyperbolic geometry two similar triangles must be congruent.*

Proof. Since they have the same corresponding angles, by Gauss' theorem they must also have the same area. But the corresponding lengths squared are proportional to the areas. Thus they must have the same corresponding lengths. □

Finally, we mention that just as in Euclidean geometry, where the separation of geodesics is linear, given any two geodesics γ_1 and γ_2 emanating from a point P in hyperbolic space and going out a distance t on each geodesic, another way of understanding what hyperbolic means is to say that $d(\gamma_1(t), \gamma_2(t))$ tends to infinity as $t \to \infty$ as e^{kt}, $k > 0$, tends to infinity.

Elliptic Geometry.

This is the geometry (for us in 2-dimensional space) in which, given a line $\mathcal{L}$ and a point P not on $\mathcal{L}$, there are no lines through P parallel to $\mathcal{L}$. As we shall see this is exactly the geometry of the airline pilot, but not of the astronaut.

Here, for a model, we shall take for the space the surface of the unit sphere in 3 dimensional Euclidean space. For point, an ordinary point on the surface and for line any "great circle", that is, the intersection of a plane passing through the center of the sphere with the surface itself. Now any two distinct points on the surface of a sphere, together with the center, determines a unique plane, which in turn determines a unique great circle passing through these points. We call two points P and Q of

S^2 antipodal if $P = -Q$. Notice that the above argument doesn't work if the two points are antipodal since the plane they determine is not unique. Also any two given lines must meet, but not in a unique point. Rather they meet in a pair of antipodal points. Now, on the surface of a sphere the *shortest arc* joining two non-antipodal points is this great circle. If the points are antipodal then there is a shortest distance but not a unique shortest path. Notice also that here there simply *are no parallel lines.* This is because any two great circles meet in a pair of antipodal points, that is some point of our space.

Clearly, ordinary Euclidean linear isometries when restricted to the sphere give elliptical isometries of it, since they take great circles into great circles and preserve their length. In fact, our next result, which follows from the section on linear algebra above, shows that $Isom(S^2)$ is isomorphic in a natural way with the orthogonal group of $\mathbb{R}^3$.

Theorem 3.2.5. *Let $L \in O(3, \mathbb{R})$. Then L leaves the unit sphere S^2 invariant and if T is the restriction of L to S^2, then T is an isometry of S^2. Conversely, each isometry T of S^2 is the restriction of a (unique) $L \in O(3, \mathbb{R})$.*

Hereafter, we shall not distinguish a linear isometry of $\mathbb{R}^3$ and its restriction to S^2. The following result due to Euler and whose proof we omit, plays a role here.

Corollary 3.2.6. *Each isometry of S^2 is a rotation about some axis, or is the product of such a rotation with a reflection.*

Corollary 3.2.7. *The isometries of S^2 coming from rotations act transitively. One that fixes a point, say the north pole, is a rotation about the axis through N. Thus $S^2 = \mathrm{SO}(3)/\mathrm{SO}(2)$.*

Proof. If P is a point on the sphere $P \neq N$, then form the plane determined by P, N and the center of the sphere. This plane intersects the sphere in a great circle (a circle of longitude) on which lie N and P. Clearly, we can pass from P to N by an appropriate rotation.

If g is a rotation fixing N, then g fixes the line $\mathcal{L}$ through N. Hence by the lemma below g stabilizes its orthogonal complement $\mathcal{W}$. Therefore g stabilizes $\mathcal{W} \cap S^2 = S^1$. Since the determinant of g is 1 it follows that g is a rotation in this plane and so $g \in \mathrm{SO}(2)$. □

Lemma 3.2.8. *If $G = \mathrm{SO}(3)$ acting on $\mathbb{R}^3$ and G fixes a line l passing through the origin then G preserves (but does not fix) the orthogonal complement $l^\perp$.*

Proof. Let $v \in l^\perp$ and $g \in G$. Then $gv \in l^\perp$. This is because $(gv, l) = (v, g^{-1}l)$ and $g^{-1}l = l$ so $(v, g^{-1}l) = (v, l) = 0$ □

Corollary 3.2.9. *The sum of the angles of any triangle is* greater *than 180 degrees.*

Proof. Suppose not. Then the sum of the angles of a given triangle is either less than or equal to 180 degrees. Suppose there were some triangle A, B, C whose angle sum is 180 degrees. Draw a geodesic γ through C making angle B with side B, C . It follows that the angle between γ and A, C also equals angle A. On the other hand, continuing both γ and A, B in each direction they must meet in angle θ_l and θ_r, respectively. But, by hypothesis, $B \geq \theta_r + A$ and $A \geq \theta_l + B$. Therefore, $B \geq \theta_r + A \geq \theta_l + \theta_r + B$, and this means $\theta_l + \theta_r \leq 0$, a contradiction. Thus the sum of the angles of all triangles is less than 180 degrees. This is equivalent to the hyperbolic parallel postulate and contradicts the elliptic situation. □

Since the situations described by the three parallel postulates are mutually exclusive and cover all possibilities, and the same is true for

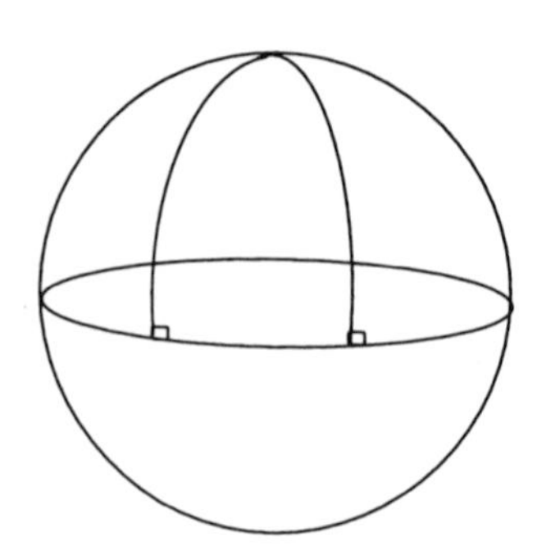

Figure 3.19

the excess of a triangle, the corollary above implies the elliptic parallel postulate. Thus the two are actually equivalent.

We remark that the theorem due to Gauss mentioned above also works in elliptic geometry, but here we must define the excess of a triangle $\mathcal{T}$ by $ex(\mathcal{T}) = (A + B + C) - \pi$. Then we have

Theorem 3.2.10. *For an elliptic triangle, its excess equals its area.*

Similarly, by the same proof as above, we get

Corollary 3.2.11. *In elliptic geometry two similar triangles must be congruent.*

Finally, we mention that just as in Euclidean geometry, where the separation of geodesics is linear, and in hyperbolic geometry where it is exponential, given two geodesics γ_1 and γ_2 emanating from a point P in elliptic space and going out a distance t on each, what happens is eventually these geodesics must meet; $\gamma_1(\frac{\pi}{2}) = \gamma_2(\frac{\pi}{2})$.

Concluding remarks:

Although it only takes two parameters to locate a point in the plane *or on the 2-sphere, or in hyperbolic 2-space*, there are other situations in which it would take more that two parameters to specify *a "point" or an "event"*. For example, one might want to simply locate a point in 3-dimensional space. A more complex situation is if one intended to meet someone at a certain point in 3-dimensional space at a certain *time* (otherwise this "event" would not occur). Evidently, such an event is

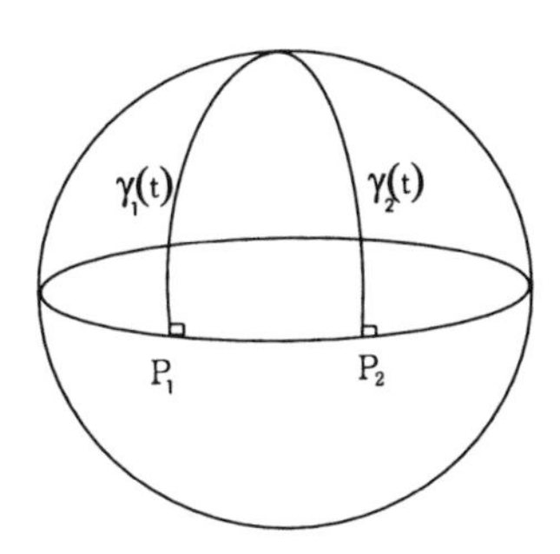

Figure 3.20

specified by giving 4 real parameters. Similarly, if one were to be at a certain point in 3 dimensional space at a certain time and seek to meet someone coming from a certain direction, or arriving at a certain speed, or wearing a certain color raincoat (which could be specified by giving a frequency or wavelength), then there would be even more parameters involved. This is why the study of higher dimensional space is really built into the study of 1-,2- or 3-dimensional space and is unavoidable.

3.3 Some Two- and Three-Dimensional Topology

Analogously to the definition of geometry, the formal definition of topology is the study of those properties, of a surface say, which remain invariant, that is are unchanged, under what are called homeomorphisms.

Heuristic Definition: A *homeomorphism* between two spaces thought of as made of rubber is any deformation of the rubber taking one to the other, but making sure the rubber doesn't tear. Two spaces which differ by a homeomorphism are called *homeomorphic* .

For example, the surface of a coffee cup is homeomorphic with the surface of a doughnut. But cannot be homeomorphic with the surface of a sphere because the whole cannot be eliminated without tearing.

Thus we have here a much larger group and, unlike what happened in the previous sections on geometry, distances and angles will definitely be changed. But remarkably enough, certain things of importance remain unchanged. If one had a sphere with a handle attached , this could be manipulated to create the surface of a doughnut. But it would always have a hole and therefore could never be distorted to the surface of the sphere itself. Thus, topology studies objects on a more general or abstract level than say geometry.

We begin with the notion of *polyhedron* in 2- or 3-dimensional space. A polyhedron is any connected geometric figure composed of points, called vertices with certain of them connected by line segments, called edges, and (in 3-dimensions) some of the planar facets filled in (just as the edges fill in between vertices) by what we shall call faces. Now, a distortion of a polyhedron which makes the edges and surfaces curved has no significant effect on the combinatorial properties of such an object. For example, it certainly doesn't affect the number of vertices, edges or faces. As we shall see in a moment, it will be convenient to consider such a "curvilinear" polyhedron also to be a polyhedron.

For a polyhedron S we shall call the alternating sum of the number of its vertices, edges and faces its Euler characteristic.

$$\chi(S) = V - E + F.$$

Here V, E, and F are respectively, the number of vertices, edges and faces of the polyhedron. Now, any closed surface S can be triangulated, i.e., broken up into a polyhedron like figure, but perhaps with curved edges and faces, and on that basis we can calculate $\chi(S)$. We shall see that this integer is the same no matter how we construct the break up. $\chi(S)$ depends only, or perhaps it is better to say intrinsically, on the surface S. We shall largely restrict our attention to the case of the sphere and make a few remarks about generalities at the end. Assuming what we have just claimed, we can therefore calculate the Euler characteristic, $\chi(S^2)$, of the 2-sphere by considering a tetrahedron which is clearly homeomorphic to S^2. Thus $\chi(S^2) = 4 - 6 + 4 = 2$. We now turn to the proof of the invariance of the Euler characteristic of the 2-sphere.

Theorem 3.3.1. $\chi(S^2) = 2$

Proof. To prove the invariance of the Euler characteristic for a sphere, first fill in any of the polygonal faces which happen to have more than 3 sides by additional edges to make each of them into triangles ; if the face has n sides then we get $n - 2$ triangles. But notice that with each of these operations, although we have increased the number of faces by $n - 2 - 1 = n - 3$, we have left the number of vertices the same and we have also increased the number of edges by $n - 3$. Thus the Euler characteristic has not been changed. We may therefore consider the polyhedron to have faces consisting only of triangles. Next, place one of the faces of the polyhedron on a table and remove it. If we can show for this modified polyhedron that $\chi = 1$, then we will know $\chi(S^2) = 2$. Now blow this up so that when its shadow is projected onto the table,

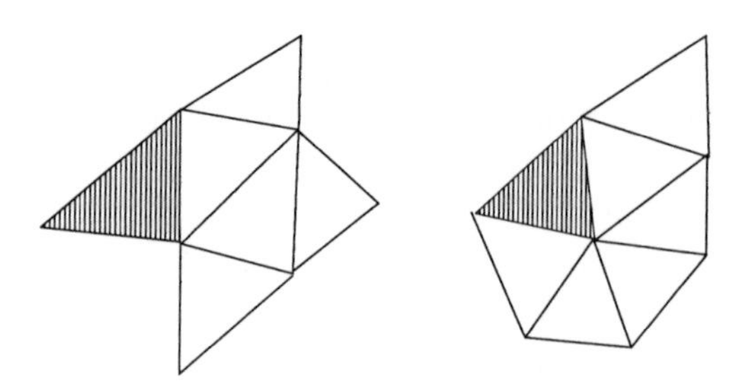

Figure 3.21

say from an overhead light, no two edges intersect. Notice that under this transformation all metric properties such distance, angle and area have been distorted (changed), but the basic combinatorial features of the network of vertices, edges and faces has been preserved. Thus the combinatorial topology has not been changed.

Starting from the outside remove a triangle. There are now two possibilities: Either the new figure has one less face, one less edge(see fig.3.3), and the same number of vertices as before, or it has two less edges, one less face and one less vertex. In either case it has the same Euler characteristic as before. Continuing in this way we get down to a single triangle, which clearly has $\chi = 1$. □

Now, one may ask, which aspects of a surface S is important in determining its Euler characteristic? It is a theorem of 2-dimensional topology that a closed surface of the type we are thinking of here is the 2-sphere with a certain finite number g of handles attached. This g in an invariant, called the genus, of the surface. The 2-sphere itself is considered to be a sphere with no handles attached. Thus its genus is zero. Another example is the surface of a doughnut (or, for New Yorkers a bagel). This is clearly a sphere with one handle; thus its genus is 1. It is a further theorem of topology that for any such surface of genus g we have $\chi(S) = 2 - 2g$. Notice that this agrees with our calculation above: Since $g = 0$, we see that $\chi(S^2) = 2 - 2g = 2$. Notice also that, by virtue of this formula, the Euler characteristic of a doughnut is zero and in fact this can be the case only for the doughnut.

We shall now give an application of the Euler characteristic of a sphere to classifying Platonic solids. A Platonic solid S in 3 space is a polyhedron that is as symmetric as it can possibly be. Each of its faces is congruent, the number of edges emanating from each vertex is constant (independent of the vertex) and angles which those successive edges form are all equal. Notice that in the plane there are infinitely many of these, one for each integer $n \geq 3$. Now, in addition to the definitions of V, E, and F given above, for any polyhedron, in the case of a Platonic solid, we also denote by n, the number of edges on each face and r, the number of edges emanating from each vertex. Clearly, the integers n and $r \geq 3$. A Platonic solid is determined, up to similarity, by n and r. We shall now investigate what the real, as opposed to theoretical, possibilities are for n and r.

Now, since each edge is actually on two adjacent faces and each face has n edges, we see that $nF = 2E$. Similarly, since each edge connects two vertices and there are r edges emanating from each of V vertices, we see that $rV = 2E$. In all cases, since the Platonic solid is homeomorphic with the sphere, we know $V - E + F = 2$. Thus $\frac{2E}{r} - E + \frac{2E}{n} = 2$. Dividing by $2E$ yields

$$\frac{1}{r} + \frac{1}{n} = \frac{1}{2} + \frac{1}{E}.$$

Since $\frac{1}{E} > 0$, it follows that $\frac{1}{r} + \frac{1}{n} > \frac{1}{2}$. Therefore, if n and r are both ≥ 4 we get a contradiction. Thus $r = 3$ or $n = 3$, or both. If $r = 3$, then by the equation above n must be 3, 4 or 5. Similarly, if $n = 3$, then r must be 3, 4 or 5. Throwing together all these possibilities and coalescing any that are the same we see, by the above, that there are exactly five distinct possibilities. Namely,

1. $r = 3$ and $n = 3$, in which case $V = 4$, $E = 6$ and $F = 4$. Here S is a tetrahedron.

2. $r = 3$ and $n = 4$, in which case $V = 8$, $E = 12$ and $F = 6$. Here S is a cube.

3. $r = 3$ and $n = 5$, in which case $V = 20$, $E = 30$ and $F = 12$. Here S is an icosahedron.

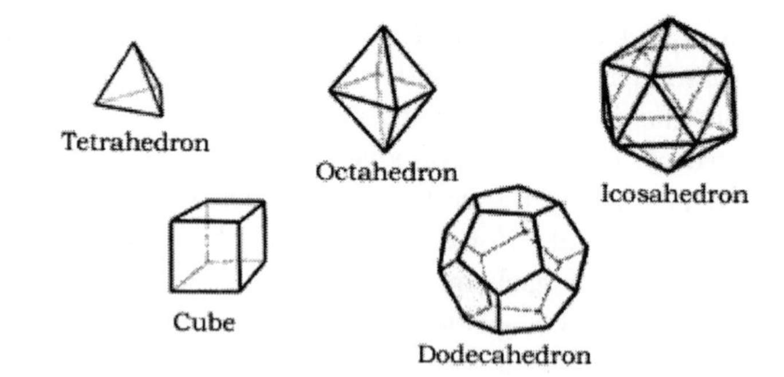

4. $n = 3$ and $r = 4$, in which case $V = 12$, $E = 30$ and $F = 20$. Here S is a dodecahedron.

5. $n = 3$ and $r = 5$, in which case $V = 6$, $E = 12$ and $F = 8$. Here S is an octahedron.

Notice that there is some kind of duality among the Platonic solids given by considering those with the same number of edges and interchanging the numbers of vertices and faces. Interchanging the numbers of vertices and faces in the cube gives the octahedron and vice versa, while doing this to the icosahedron gives the dodecahedron and viceversa. The tetrahedron is self dual.

An important principle, which we will make much use of in the remainder of this chapter, is the following, which is intuitively obvious and actually not difficult to prove once one has set forth the right definitions.

Deformation principle: Suppose two geometric, or topological structures X and Y each have a topological invariant $n(X)$ and $n(Y)$, respectively, where these invariants are integers. If X and Y can be continuously deformed into one another, then $n(X) = n(Y)$. For if $n(X) \neq n(Y)$, then during the course of the deformation $n(X)$ goes to $n(Y)$ continuously, but cannot do so without passing through non integer values, a contradiction.

We now return to the Fundamental Theorem of Algebra mentioned in chapter 1. It is interesting that it seems to take methods completely out of the realm of algebra to prove this theorem of algebra.

Theorem 3.3.2. *A polynomial equation* $p(z) = a_n z^n + a_{n-1} z^{n-1} + \ldots + a_0 = 0$*, with complex coefficients* a_i*, of* positive *degree* n*, must have a complex root. In fact, it has all* n *roots, all in* $\mathbb{C}$*.*

Proof. By the division algorithm and induction on the degree n of $p(z)$ we need only show that $p(z)$ has some complex root. Also, there is no harm in assuming that $a_n = 1$. Suppose $p(z)$ is never zero. If z traces out some closed curve once counterclockwise in $\mathbb{C}$, $p(z)$ will trace out some closed curve in $\mathbb{C}$ which will *never pass through* O, the origin. We define the *index* of this curve with respect to O to be the number of complete revolutions about O, either clockwise or counterclockwise that $p(z)$ traces out as z runs around the given curve once counterclockwise.

For example, we can take for the curve a circle centered at O of radius $r \geq 0$. In this case the index is clearly an integer valued continuous function $\text{ind}_p(r)$ of r. Notice that, although trivial when $r = 0$, this makes sense here and $\text{ind}_p(0)$ is clearly 0. We will show that for r large enough $\text{ind}_p(r) = n$. Then, since this is a continuous function of r taking integer values, and by the deformation principle it cannot go from 0 to n continuously without passing through non integer values, the only possible conclusion is $n = 0$, a contradiction.

To see that for large r, $\text{ind}_p(r) = n$, first consider the simpler problem of proving this for $f(z) = z^n$. This is obvious from De Moivre's theorem and, in fact, doesn't even require r to be large, just positive. For the general case first note that for $|z|$ larger than all the norms $|a_i|$ of all the coefficients of p, we will show that $|p(z) - z^n| < |z|^n$. This follows from the triangle inequality (in fact we have essentially seen this argument before)

$$|p(z) - z^n| \leq \Sigma_{i=0}^{n-1} |a_i| \cdot |z|^i = |z|^{n-1} \cdot (\Sigma_{i=0}^{n-1} |a_i| \cdot |z|^{-i}) < |z|^n.$$

This inequality means that for $r = |z|$ sufficiently large, the distance between $p(z)$ and z^n is less than the distance between z^n and O. It follows that if z is on the circle of radius r centered at O, the straight line segment l_z, joining $p(z)$ and z^n, cannot pass through O . Hence we can continuously deform the curve traced out by $p(z)$ into the curve traced out by z^n, without ever passing through O, simply by pushing $p(z)$ along the line segment l_z to z for each z of modulus r. Taking r fixed and sufficiently large, since p can be deformed into f continuously, avoiding the origin, and both $\text{ind}_p(r)$ and $\text{ind}_f(r)$ take only integer values, they must be equal. Thus $\text{ind}_p(r) = n$, for large r. □

Finally, by similar techniques, we prove another important fact of two dimensional topology known as the Brouwer fixed point theorem.

Let D denote the closed unit disk in the plane, that is D consists of points of the interior *together with those on the boundary.* If f is a map of D to itself we call a point x of D a *fixed point* of f, if $f(x) = x$. The reader should think about whether the statement of fixed point theorem below would be true if we considered D *without* $\partial(D)$, the boundary or circumference. Of course, by composing appropriate maps the fixed point theorem also holds for any space homeomorphic with D. This is in spite of the fact that the argument we give depends crucially on the boundary being smooth and, in many cases, a space homeomorphic with D can have a non smooth boundary, e.g. a triangle. The theorem is also true, and of considerable importance in higher dimensions, but more machinery is needed to prove it there.

Theorem 3.3.3. *Let $f : D \to D$ be a continuous map of D to itself. Then f has a fixed point in D.*

Proof. If not, then $f(p) \neq p$ for all points $p \in D$. For each p draw an arrow from p to $f(p)$. This mass of non-zero arrows, one at each point of D, with the arrows varying continuously from point to point, is called a *continuous vector field* on D *without singularities.*

Let us now restrict our attention to the points p on the boundary and the arrows emanating from these points. Choose a particular point, say p_0, on the boundary and go around the boundary once in a counterclockwise direction observing these arrows. When we return to p_0 the arrow emanating from it will be the original one. Hence the arrows may have wound around p_0 any integral number of times clockwise or counterclockwise (including possibly zero times), depending on how many times and in which direction we assign positive, zero or negative values. Thus we have associated an *index*, or winding number, to this mass of arrows of $\partial(D)$ with respect to the point p_0. We call this integer, $\text{ind}_{p_0} = 0, \pm 1, \pm 2 \ldots$, the index of the vector field at p_0. We now will show $\text{ind}_{p_0} = 1$.

Lemma 3.3.4. $\text{ind}_{p_0} = 1$.

Proof. Suppose $\text{ind}_{p_0} = n$ where, for example, $n \geq 2$. Consider the unit tangent vector field $\mathbf{T}$ on the boundary going around counterclockwise. Its index is clearly 1. Starting with p_0 and going around counterclockwise, the arrows of the original vector field must pass $\mathbf{T}$. Hence there must also be a point where the vector points in exactly the same direction as the tangent vector. This is impossible since the original vector field always points into D and therefore can't be tangential. Similarly, if $\text{ind}_{p_0} \leq 0$, there is some point where the vector points in the tangential direction. Hence $\text{ind}_{p_0} = 1$. □

Continuing the proof of the theorem, now consider a circle Γ_r centered at the origin in the interior of D of radius $0 < r < 1$ and choose any point $p_1(r)$ on it. Then, by deforming this circle to the boundary and taking $p_1(r)$ to p_0, by a rotation for example, we see by the deformation principle that $\text{ind}(\Gamma_r)_{p_1(r)} = 1$. This is true no matter how small r is. But by continuity of the vector field at the origin there is a sufficiently small subdisk about the origin in which all vectors point in essentially the same direction as the one emanating from the origin, say to within 5 degrees. For this r $\text{ind}(\Gamma_r)_{p_1(r)} = 0$, a contradiction. This proves the theorem. □

The following result means that if D were made of rubber, unless you tear the membrane you can't map it onto the boundary

Corollary 3.3.5. *There is no continuous map $f : D \to \partial(D)$, which leaves the boundary fixed.*

Proof. Suppose f were such a map. Let $I : \partial(D) \to D$ be the injection of $\partial(D)$ into D, and $g : D \to D$ be the continuous map which reflects each point of D through the origin. Then since g preserves the boundary, the composition $g \cdot I \cdot f$ is a continuous map of D to itself which clearly has no fixed points, a contradiction. □

We remark that Theorem 3.3.3 and Corollary 3.3.5 actually hold in $\mathbb{R}^n$. Since we were speaking of vector fields in the course of proof of the Brouwer fixed point theorem, it might be worth mentioning that there can be no *tangential* vector field without singularities on the 2-sphere, S^2, but there is one on the doughnut. With regard to such vector fields, the reader might find it interesting to draw a vector field on the sphere with two singularities of index 1 each, as well as a vector field on the sphere with only one singularity, but of index two. One should observe that on the doughnut there are two linearly independent tangential vector fields without singularities. Can you now make a conjecture about the existence or non-existence of such vector fields in terms of the Euler characteristic of the surface? If you can not, consider our final result (due to Poincaré) whose proof we omit.

Theorem 3.3.6. *If S is a 2-dimensional surface on which is a continuous vector field with a finite number of singularities at say $\{p_1, p_2, \dots p_n\}$, then $\Sigma_{i=1}^n \operatorname{ind}_{p_i} = \chi(S)$. In particular, if S has a vector field without singularities, then $\chi(S) = 0$.*

Since, as we proved above, if $\chi(S) = 0$ ie $g = 1$ and S is a doughnut, then S has a continuous vector field without singularities. (In fact, it is not difficult to see that S has two linearly independent continuous tangent vector fields without singularities). Thus, combining this with the theorem of Hopf, we see that a surface S has a vector field without singularities if and only if $\chi(S) = 0$. In particular, there is no continuous vector field without singularities on the sphere.

Index